The Fragmented Forest

The Fragmented Forest

Island Biogeography Theory and the Preservation of Biotic Diversity

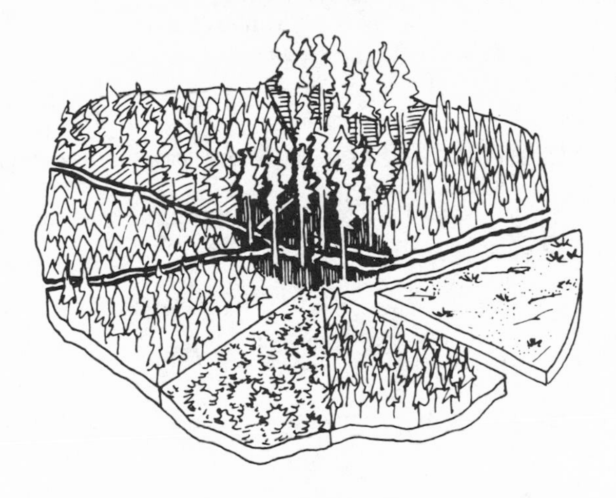

Larry D. Harris

With a Foreword by Kenton R. Miller

The University of Chicago Press

Chicago and London

The University of Chicago Press, Chicago 60637
The University of Chicago Press, Ltd., London

Printed in the United States of America
93 92 91 5

Library of Congress Cataloging in Publication Data
Harris, Larry D.
The fragmented forest.

Bibliography: p.
Includes index.
1. Nature conservation. 2. Biogeography.
3. Island ecology. I. Title.
QH75.H37 1984 639.9 84-144
ISBN 0-226-31763-3
ISBN 0-226-31764-1 (pbk.)

To

CURTIS GEORGE HARRIS

who came to appreciate that natural resources are not given to us by our fathers but are loaned to us by our children

Contents

Part 3: Analysis of Alternatives

Part 4: A Planning Strategy

Foreword

Present and predicted rates of extinction of plant and animal species have dramatic implications for all of us. Students of this problem have documented the economic values associated with species in terms of wild relatives of crop cultivars, timber trees, pharmaceuticals, industrial chemicals and other materials. The ethical and spiritual values involved are also of the highest importance.

The tropical rain forests of the world harbor the majority of the planet's species, yet this wealth of species is being quickly spent. While the exact numbers of species involved and the rate of forest clearing are still under debate, the trend is unmistakable—the richest terrestrial biome is being altered at a scale unparalleled in geologic history.

More complex scientific aspects of the problem have been explored recently by Michael Soulé, Bruce Wilcox, Sir Otto Frankel, Peter Raven, Christine Schoenwald-Cox, and others. Growing awareness of the importance of the problems of extinction and the loss of biological diversity has prompted accelerated activities and investments in conservation worldwide. Symposia, workshops, classes, research, and the scientific literature have contributed to the debate. Public policy forums have included such events as the United States Strategy Conference on Biological Diversity.

In practice, the preservation of wild plant and animal species has generally been carried out through the establishment and management of national parks and other types of protective nature reserves. Over 120 nations have now established some 3,100 national parks and similar conservation units, totaling over 400 million hectares.

With this much land dedicated to nature conservation, can we sit back with the confidence that all or even most species have a reasonable chance of surviving? Ecological science warns that while nature reserves are critical to ensure the maintenance of plant and animal species, they are far from sufficient to meet the needs of species survival. The maintenance of biological diversity requires special measures that extend far beyond the establishment of nature reserves.

Several reasons for this stand out. Existing reserves have been selected according to a number of criteria, including the desire to protect nature, scenery, and watersheds, and to promote cultural values and recreational opportunities. The actual requirements of individual species, populations, and communities have seldom been known, nor has the available information always been employed in site selection and planning for nature reserves. The use of lands surrounding nature reserves has typically been inimical to conservation, since it has usually involved heavy use of pesticides, industrial development, and the presence of human settlements in which fire, hunting, and firewood gathering feature as elements of the local economy.

Furthermore, existing nature reserves have often inadvertently become parts of landscapes displaying the pathos of poverty, underdevelopment, and social inequity. Under such conditions, rural peoples find themselves obliged to forage in the reserves for food and energy and basic building materials. The very concept of a reserve is seen as opposed to human welfare, with the conservation movement unwittingly ending up on the side of "antidevelopment," and ultimately being seen as "antipeople."

But the tide is changing. Without being overoptimistic, it is fair to conclude that present work on nature reserves has begun to focus upon a more social perspective for field action. At the 1982 Third World Congress on National Parks, protected-area planners, managers, directors, policymakers, and conservationists examined the status of reserves worldwide and presented experience from the past decade in which "sustainable" approaches to park planning and management have been implemented. Recent developments include the following:

1. Recently established parks are being incorporated into regional resource management schemes.
2. Parks are being surrounded by zones of compatible land use.
3. Communities in adjacent lands are deriving benefits from parks management.
4. Although protective management continues to be a major goal of the parks, the boundaries are becoming less rigid, the local community is becoming less hostile, and the concept of "benefits to

the people" is being applied not only to future generations, but also to those already present.

Generations to come will be grateful to those who set up the network of national parks and nature reserves around the world. The biggest threat to these accomplishments, however, is that, looked at in terms of areas, millions of hectares, and improvements in social compatability, they could lull us into believing that the job is well in hand.

Recent studies demonstrate that most existing protected areas are small, have odd shapes, and are at considerable distances from one another. Few have species lists. Most have no research program. Research on the effective boundary for conservation purposes versus actual legal boundaries has only recently been initiated. The fact is that most national parks and nature reserves will rapidly become green islands surrounded by agriculture, logging operations, urbanization, and encroaching desert.

What then is the alternative? This book offers an approach that holds great promise. Larry Harris employs the principles of island biogeography and other aspects of biological and ecological science to provide a set of guidelines for rural planning. More specifically, his approach treats patches of old-growth forest as islands in a sea of tree plantations or human-dominated landscape. He details a scheme for surrounding each patch with a low-intensity forest management buffer zone (long-rotation management) and then considers how these long-rotation islands should function as a system, an archipelago. He reasons that many (if not most) species of wildlife will not be secure within any single patch and therefore movement between patches must be anticipated and planned for. Principles of contour, topography, and energy transformation are invoked to fit the island system to the landscape. The scheme that emerges thus integrates a conservation strategy into a developing or developed landscape. The fieldwork and data behind this study concentrate upon the Pacific Northwest of the United States. Given the existing stands in the Douglas fir region and the extensive network of national forests and national parks, the approach promises to be widely applicable.

The particular relationship of this approach with the United States concept of the national forest deserves particular mention. The national forest provides for public ownership of large tracts of forest land. Within the individual forest unit, management is planned and practiced through zoning that allows for the staggering of harvested areas, the protection of watersheds and fragile areas, the closure of key natural areas or research sites to harvesting, and many types of activity besides the removal of timber, including recreation, hunting, fishing, and grazing. Harvesting in national forests, which are obviously distinct from national parks and related protective management areas, is carried out in such a manner as to maintain all species and the productive capacity of the land.

Thus the national forest as a management category provides for both total protection and harvesting. This managerial concept is relatively clear. Forestry practice, however, has not always been clear. The central issue of this study is how to retain stands of old growth while harvesting timber as necessary and appropriate. The old growth is required to maintain certain species of plants and animals not found in earlier seral stages of forest growth. The solution to the problem is a mosaic of old-growth islands.

The global implications of the archipelago approach are important. Most wild lands of the world have been placed under the responsibility of foresters and forestry organizations, public and private. The hard products to be derived from timber, water resources, and grazing, as well as the environmental services of the forest, are all in increasing demand. Ironically, increasing requirements for goods and services from the forest are challenged by the accelerating destruction of the forest ecosystem.

The role of foresters and conservation biologists is crucial. What remains to be examined is how the proposed forestry management system can be of use in tropical and subtropical forest areas. The basic concepts involved will certainly be of interest and warrant a major trial.

The archipelago approach is a welcome contribution to the global effort toward the maintenance of biological diversity. It is entirely consistent with the World Conservation Strategy. The principles suggested are relevant for foresters, natural resources managers, and town and country planners everywhere.

Kenton R. Miller
Director General
International Union for the Conservation
of Nature and Natural Resources

Preface

It is not often that the conjunction of problem, people, and events leads to the development of significant contributions toward the solution of the problem. Yet just such a conjunction occurred in the northwestern United States in 1980. By that time previously uncut, old-growth forests had been virtually eliminated from state and private lands, leaving the old-growth forests on federal lands as the principal source of large-size-class timber. Removal of this timber from federal lands had increased exponentially over the previous few decades, causing concern, and in some cases alarm, among groups and organizations dedicated to the preservation of environmental quality. For those with an acute awareness of the rapid loss of species and genetic diversity, the management of the forests of southwestern Oregon became as critical as management of the rain forests of northwestern Brazil. A complicating factor was that there was not unanimity of opinion on how the old-growth forest should be managed. While one side feared the environmental consequences of liquidation and fragmentation of the old-growth forest, others feared the economic consequences of slowing the removal. It was speculated that reduced sales of timber, even at standard bid prices that were only a fraction of the cost to replant and regrow, would wreak havoc in counties and states economically dependent on the timber industry.

Several scientific and statutory developments also entered into the situation. Under the sponsorship of the U.S. International Biological Program (IBP), large amounts of data about the old-growth Douglas fir ecosystem had been gathered. These data made clear that old-growth forest ecosystems had an importance to human society and to the biosphere far in excess of their value as wood fiber. This awareness contributed in part to the passage of the 1976 National Forest Management Act, which mandates a comprehensive planning approach to multiple-use management and the preservation of biotic diversity on national forest lands. By 1980 the planning process was in full swing but the first plan had not yet been developed. Thus, although a great deal of scientific information was available, it was not in a form readily usable for comprehensive planning, nor was it clear that the scheduling of timber operations in hundreds of districts and thousands of old-growth tracts would be greatly or immediately improved by increased knowledge of the internal functioning of the old-growth ecosystem.

A possible resolution to this impasse was provided by island biogeography theory, which was developed in the late 1960s. This theory caught the imagination of many scientists, and its utility for resource management was being actively investigated. It was widely believed that explanations for the distribution and abundance of organisms on true islands might also be applicable to forest habitat islands surrounded and isolated by clearcuts, regeneration stands, tree plantations, or humanized environments. If this were so, then the predicted consequences of insularizing old-growth ecosystems could be used in comprehensive planning for the preservation of biotic diversity.

The purpose of the present work is to draw together available scientific information from the western Cascades and use it to evaluate the utility of island biogeography theory as a guide to comprehensive planning for the conservation of old-growth ecosystems in the context of managed forest lands.

But before principles and procedures can be accepted as applying generally, they must be shown to apply to at least some specific areas. Spanning nearly 500 miles of latitude, the Douglas fir forest of the western Cascades remains the largest continuous tract of uncut forest in the lower United States. This large expanse of unique forest, with characteristics of a temperate rain forest, provided an ideal testing ground. Having worked in the 30 million acres of managed commercial pineland in the southeastern United States for ten years, I concluded that a future forest by design would be superior to any we might inherit by default. This conclusion led directly to the challenge of transforming the implications of island biogeography into forest management applications and to the integration of conservation planning into development planning. To counter the criticism that data and principles developed in the eastern United States might not necessarily apply in the western Cascades, I purposely limited my citations to western studies. This has meant, unfortunately, that the work of several leading scholars has not been cited; it has not, however, been ignored.

Because of my attempt to keep the arguments data-based, most of the discussion centers on vertebrates; few if any arguments are based on invertebrates. In the course of analyzing Douglas fir forest data, it became clear that, among the vertebrates, the overwhelming dominance of carnivores is one of the most striking faunal characteristics. The carnivores are all wide-ranging creatures, most of which have very low densities. Wolverine and lynx are very rare, and grizzly bear, gray wolf, and fisher have already been extirpated from western Oregon. Because these species cannot be restricted to any single old-growth patch, it is impossible for any single stand of old growth to "contain" a complete faunal assemblage. Given that it is important to conserve top carnivore species and that these animals probably perform important biological functions in natural old-growth ecosystems, the planning strategy had to be aimed at a level that would maintain their presence and influence. For various reasons, the emphasis had to be shifted from the old-growth system to the system of old growth. I refer to this as the island archipelago approach. Simple physiographic and energy principles are used to tie the island system to the landscape.

The University of Florida granted me faculty development leave, and Arnett C. Mace, Jr., director of the School of Forest Resources and Conservation, handled all the administrative details involved. Robert Ethington, director of the Pacific Northwest Forest and Range Experiment Station, U.S. Forest Service, worked out logistic details in Oregon. Dave Luman and Bill Neitro of the Bureau of Land Management (BLM) provided essential agency endorsement for funding. Jerry Franklin, Jack Thomas, and Chris Maser provided critical scientific advice, and Chris Maser contributed many hours of valuable discussion time. Art McKee introduced me to the H. J. Andrews Forest and helped me detail recent trends in the cutting of old growth. His depth and breadth of knowledge of Cascades forest ecology was invaluable. Discussions with Scott Overton and Benee Swindel, both statisticians, frequently clarified issues and convinced me that finding the right question is often more difficult than finding the correct answer. Bob Giles, Don Grayson, Jack Thomas, Dick Taber, Henry Gholz, and several other reviewers provided important criticism and support. Ralph

Jaszkowski, Bill Emmingham, Charlie Phillips, and Al Lang helped me distinguish between reasonable forest management prescriptions and overly idealistic principles. Bob Frenkel and Sarah Greene provided information on Research Natural Areas and drew figure 6.14. Dave Maehr and Joyce Lottinville provided illustrations, while Andreas Richter, Lynn Badger, and Brian O'Kelly worked on analytical details and helped with editing the manuscript. Sebrina Street and Pat Linihan typed several editions of the text and made personal sacrifices beyond the call of duty. I owe a debt to all these people.

PART

1

Problem Setting

1

Introduction

Recent changes in the forests of the world have received attention at the highest scientific and governmental levels (I.U.C.N. 1980; C.E.Q. 1980; Meyers 1980; Lanly 1982; U.S.D.A. 1982). The most notable changes fall into three principal categories: (1) reduction in total forest acreage; (2) conversion of naturally structured and regenerated forests to even-aged monoculture plantations; and (3) fragmentation of remaining natural forests into progressively smaller patches isolated by plantations or by agricultural, industrial, or urban development (Harris 1980). These processes are not independent of one another and must be considered in an interrelated fashion.

Worldwide wood consumption has increased to approximately 100 billion cubic feet per year (F.A.O. statistics in U.S.D.A. 1982, p. 81), about half of which is used for industrial products and half for fuelwood.[1] In most less developed countries, 90% of the people depend on firewood as their chief source of fuel (Eckholm 1980) and 90% of the wood harvested is used for energy (Meadows 1981). Since worldwide demand for wood is far outstripping supply, quips such as "it now costs more to heat the cooking pot than to fill it" are commonplace. Forests that covered approximately one-fourth of the earth's land area in 1960 now cover only about one-fifth of the land area, and will likely cover only one-sixth by the year 2000. This decline is projected to stabilize at about one-seventh of the land area by the year 2020 (C.E.Q. 1980, 117).

In spite of these trends, standing timber volumes have increased from their lowest level in the 1930s in several Western countries such as the U.S., but this is principally because of aggressive coniferous reforestation, timber stand improvement programs, and con-

1. Despite a preference for metric units, the use of English units is necessitated by their importance in, and my present reliance on, the historical literature.

version of understocked second-growth stands to tree plantations. Contemporary commercial forestry is both silviculturally and capital-intensive. Short rotation, even-age tree plantations are not like natural forests in that they are oriented toward maximum timber production, are not autogenic, and generally rely on energy-intensive management such as site modification, control of competing vegetation, genetic improvement, planting, and fertilization (Demoulin et al. 1976; Farnum et al. 1983). While the productivity of the U.S. commercial forest land has increased nearly threefold over the last twenty years, capital investments for processing have increased over sixfold (Bingham 1976; Clawson 1976).

Forest fragmentation results from patchwork conversion and development of the most accessible and/or more productive sites, leaving the remaining forest in stands of varying size and degrees of isolation (Burgess and Sharpe 1981). These forest fragments take on characteristics of habitat islands in proportion to their degree of and length of time since isolation.

Along with our increased awareness of these trends has come a worldwide concern for the preservation of biotic diversity (T.N.C. 1975; C.E.Q. 1980; I.U.C.N. 1980; U.S. Dept. State 1982). Strategy has shifted from a former species-by-species approach toward a planning approach. Thus the U.S. Department of Agriculture policy states that habitat management for viable populations "will be accomplished through the forest planning process . . ." (U.S.D.A. secretary's memorandum 9500–3, 1982). This begs the question of whether ecological theory and resource management principle can contribute to and guide the planning process.

A large and significant body of information, principle, and theory falls under the heading of island biogeography theory (sensu MacArthur and Wilson 1967). Observations and generalizations about the patterns of floral and faunal communities on oceanic islands predate the voyages of Darwin and Wallace but have only been synthesized into a cohesive body in the last thirty years (Darlington 1957; MacArthur and Wilson 1967; Sauer 1969; Pielou 1979). Conservationists have been particularly receptive to these principles because islands present such compelling conservation issues. For example, although islands represent less than 7% of global land area and less than 10% of all bird species occur there, more than 90% of all bird extinctions have occurred on islands. Of the 396 bird species listed in the International Union for the Conservation of Nature and Natural Resources (I.U.C.N.) *Red Data*

Book as threatened or endangered, 236 (60%) are endemic to islands (Gosnell 1976). If the biological principles and generalities from true islands are applicable to patches of old growth in the forest landscape, the biological foundations of a planning strategy can be established. Ecological theory might again contribute to forest management on a par with technological mechanics and economic calculus. Although analyses of conditions have been performed (e.g., Burgess and Sharpe 1981), I am not aware of any specific application of island biogeography principles to a forest management problem.

Forests of the Douglas fir[2] region of western North America represent an ideal opportunity to apply such a strategy. As the only remaining expanse of virgin forest in the conterminous United States, the old-growth Douglas fir represents a unique challenge to our planning ability and commitment to conservation. The National Forest Management Act (NFMA) and related legislation mandates that annual sales must be less than or equal to the quantity that can be removed in perpetuity. This concept, referred to as "nondeclining even flow," is at the root of the current challenge and public debate. Since old-growth forests have a low to negligible mean annual increment of usable wood (MAI), almost any harvest is above the level sustainable by that particular stand or forest type. Because present regional harvest rates are well above annual production in the Pacific Northwest, some argue that the rate of old-growth cutting must be drastically reduced to achieve a balance with production. On the other hand, young managed stands exhibit a very high mean annual increment and thus the "supply-side" view is that in order to bring production into balance with present cutting levels, the conversion of decadent old growth to young, well-stocked stands should be accelerated. One school wishes to reduce cutting down to a level equal to present production rates; the other wishes to continue or perhaps increase the conversion rate of old growth and thereby greatly increase the annual sustained yield.

Economics are also involved. Old-growth forests that have low and sometimes even negative timber growth rates represent large amounts of nearly idle capital investment. Clawson (1976) calculates that at modest interest rates, a $12 billion excess inventory in standing timber means an annual cost of $600 million in total or about $3.00 for each citizen. He asks what would result if every

2. All scientific names are given in appendix 1.

citizen of the United States were asked to contribute $3 annually toward the maintenance of this "excess inventory" of old trees. Thus, while we possess a rare opportunity to maintain significant acreages of a unique North American forest type, we also incur significant pressure to liquidate it.

To the extent that old-growth forests represent more than wood fiber and timber, we should question the trade of old growth for young growth. To the extent that remaining old-growth Douglas fir ecosystems possess unique structural and functional characteristics distinct from surrounding managed forests, the analogy between forest habitat islands and oceanic islands applies. Forest planning decision variables such as total acreage to be maintained, patch size frequency distribution, spatial distribution of patches, specific locations, and protective measures all need to be addressed. Island biogeography theory and the lessons learned from true-island biogeography provide a basis for developing a management strategy and addressing these specific decision variables.

A recent publication (Franklin et al. 1981) has synthesized what information was known of the structure and function of old-growth Douglas fir ecosystems in the western Cascades. A related volume (Thomas 1979) has detailed certain wildlife habitat implications of forest management and laid groundwork for stand scheduling decisions. Building on that information, this work describes important characteristics of the animal communities in Douglas fir–western hemlock forests of the western Cascades and develops guidelines for an old-growth management strategy.

At present, island biogeography theory represents little more than a set of interwoven ideas. The following specific application to old-growth management in the Douglas fir region will establish that it also represents a viable tool. The degree to which the approach is applied to forest management problems worldwide is dependent upon its demonstrated utility in specific regions such as the western Cascades. Although the approach outlined here will gain power in proportion to the validity of the analogy between old-growth patches and true islands, the overall strategy is sufficiently robust to be useful whether or not island biogeography theory is valid.

2

The Approach

To address the needs of forest management planning, this work attempts to (1) refine, articulate, and extend certain ecological principles in applied terms; (2) address the issue of diversity from the community and ecosystem level rather than the species-by-species approach; (3) describe and demonstrate the utility of a specific approach; and (4) generalize potential applications to other regions and forest types.

Both inductive and deductive approaches to science have been used to advance ecological knowledge and establish principles of use to resource management. The inductive approach (from observation of specifics to formulation of generalities) has dominated until recently. The strength of the approach is its reliance on empirical field data, but it is very time-consuming and has rarely resulted in theoretical formulations of wide-ranging power. The quicker, more powerful deductive approach involves the testing of hypotheses under specific conditions. The strength of this approach derives from the ability to test repetitively cause-and-effect relations between nonconfounded variables (e.g., bird species diversity $\propto$ foliage height diversity). The weakness is that "principles" all too often become encoded as truth before being tested and verified.

Since the theory of island biogeography has led to many falsifiable hypotheses and the existing data and natural history wisdom of scientists such as Chris Maser could be used to evaluate certain of these, this study draws heavily on the hypothetico-deductive approach (Hempel 1966). Field work was conducted during a short period in 1979, six months of 1980, and three months during 1981. I have used simple logic and algebraic models to piece together existing data with ecological principles and island biogeography theory in an attempt to develop a cohesive whole.

The principles used are generally independent of the validity of MacArthur and Wilson's equilibrium theory of island biogeography

(1963, 1967). It must also be borne in mind that the thrust of this work is not directed at a system of nature preserves, but rather a forest management strategy that depends heavily upon, and in turn conserves, the unique characteristics of old-growth ecosystems. Indeed, the habitat island approach to the maintenance of biotic diversity assumes, and is somewhat hinged upon, the existence of the preserve and wilderness area system.

Much of island biogeography, much of contemporary ecology, and much of the criticism of forest management deal with changing distributions, patterns, and interactions of floral and faunal communities. Until recently most references to diversity involved species diversity and the integrity of biological communities. Resource management has increasingly focused on the community level, and this is generally the emphasis herein. These analyses and principles are not directed merely at saving endangered species, but rather at keeping the full complement of species from becoming endangered. Once a species is identified as endangered we have little recourse but to drop to the species level and manage specifically at that level. The two approaches are highly complementary.

Present policies governing forest management planning must move quickly from general guidelines for development of forest plans to the detail of timber harvest scheduling on specific forests. It is at the level of the forest that standards must be converted into action plans, and situation analyses and proposals must involve specific forests. Several considerations led to the choice of the Willamette National Forest in western Oregon as an area of focus (see fig. 3.1, p. 12). As the Willamette is the top timber producer of the 155 national forests, harvest scheduling there is no small task. The H. J. Andrews Experimental Ecological Reserve is contained within the Willamette and thus the large data base from this site was available. In addition to being representative of the western Cascades, the administrative climate is ideal. However, to preclude an overly narrow focus, data were also obtained from other areas (e.g., the coast range), other national forests (e.g., the Siuslaw), and lands managed by other agencies (e.g., Bureau of Land Management).

PART 2

Current States of Nature

3

The Natural Forest Community

The Douglas fir–western hemlock forest west of the Cascade mountains in Washington and Oregon originally spanned a north-south distance of 500 miles (800 km) and occupied approximately 28 million acres (11,336,00 ha) (fig. 3.1, Sargent 1884; Andrews and Cowlin 1940). Approximately 90% of the forest consisted of Douglas fir with western hemlock and western red cedar occurring as subdominants. Old-growth Douglas fir, or trees in excess of 300 years old, probably occupied 50% of this area (14×10^6 acres) and constituted 75% of the volume (Andrews and Cowlin 1940; Kirkland 1946). Densest stocking and highest volumes occurred in the lower elevations and river valleys between 2,500 and 4,000 feet (762 and 1,219 m) (fig. 3.2). Although volumes on some sites were as great as 150,000–200,000 board feet (mbf) per acre (875–1165 m^3/ha), a reasonable average for old-growth stands was 65 mbf/acre (375 m^3/ha) (table 3.1; Gannett 1902; Plummer 1902; Langille et al. 1903; Andrews and Cowlin 1940; Groner 1949).

Detailed vegetation descriptions are given by Franklin (1979) and Dyrness and his associates (1974). Distinctive features are highlighted by Waring and Franklin (1979) and Franklin and Waring (1980). Specific attributes of old-growth Douglas fir ecosystems are reviewed by Franklin and his associates (1981). Ecosystem characteristics of particular relevance to the vertebrate fauna are discussed in six categories below.

High Latitude and Mediterranean Climate

The combination of high latitude (from 42° N to 49° N) and generally high altitude in the western Cascades creates environmental conditions ideally suited for conifers. For example, a site at 10,000 feet (3,050 m) elevation and 45° N latitude has a temperature regime approximately equivalent to that of 16,000 feet (4,900 m)

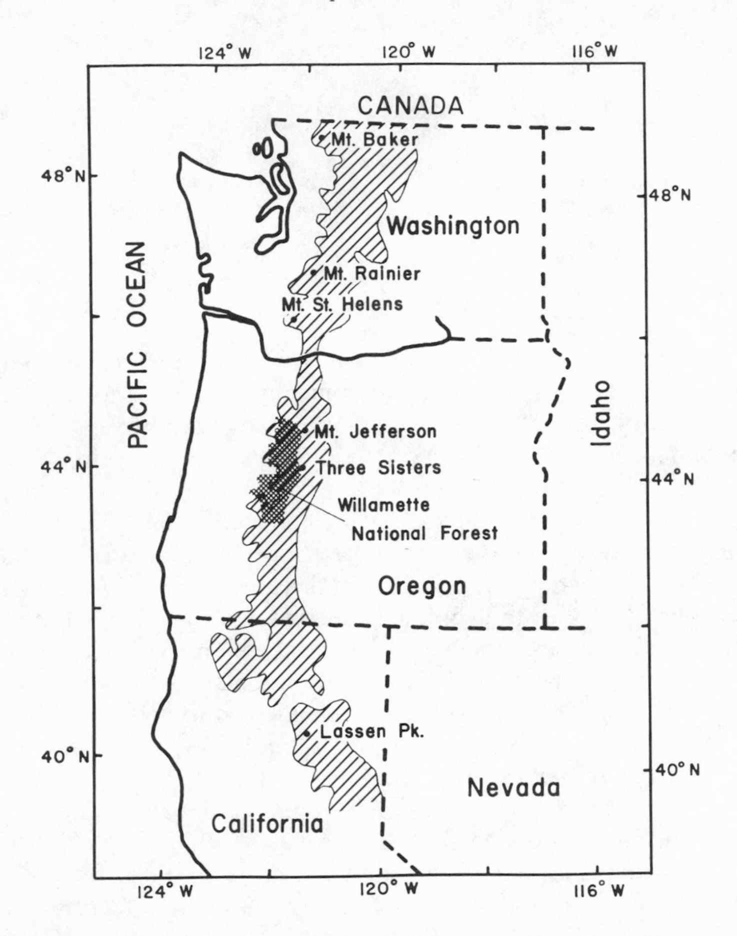

Figure 3.1 Location of Willamette National Forest in the western Cascades of Oregon.

elevation at 25° N latitude (fig. 3.3). A rainfall pattern that yields 90% of the annual precipitation during the winter and 10% during the summer (fig. 3.4) combines with the latitude and elevation to produce an environmental regime strikingly different from any other in North America. The combined effects favor conifers over hardwoods except at low elevations or on wet sites near permanent water. With little precipitation occurring during the summer, the forest quickly becomes dry and flammable, accentuating the threat and impact of fire. Forests near lakes and perennial streams take on

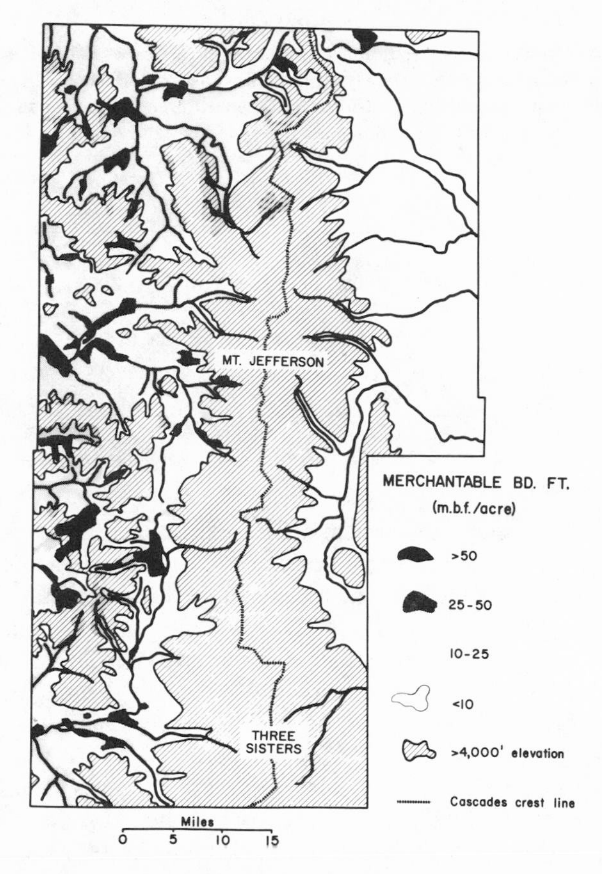

Figure 3.2 Distribution in 1901 of old-growth Douglas fir in the Willamette National Forest illustrating the location of high-volume sites at lower elevations and in river valleys (m.b.f. = 1,000 board feed gross volume). (From Harris et al. 1982, based on data from Langille et al. 1903.)

additional value as wildlife habitat because of the restricted location of free water during the reproductive period, which is also the hottest season of the year.

Canopy Height and Massivity of Forest

The most striking feature of the old-growth Douglas fir forests to the casual observer is the height and massiveness of the dominant trees (fig. 3.5). Canopy heights of mature stands average over 200

TABLE 3.1
Amount of timber acreage and timber volume in the Cascades Forest Reserve by density classes as of 1901. Data from Langille et al. 1903. (It should be noted that cruise data from this period were based on different methods and standards and yield conservative estimates by today's standards.)

Density (mbf/acre)	Area (acres)	% of area	Volume (mbf)	% of volume
0–2	778,593	46	667,567	6
2–5	382,723	23	1,339,530	13
5–10	199,022	12	1,492,665	14
10–25	263,670	16	4,614,225	44
25–50	58,772	3	2,203,950	21
50–100	1,770	—	132,750	1
	1,684,550		10,450,687	

feet (60 m) and many reach nearly 275 feet (85 m), about twice as tall as the white pine forests of northern North America and three times as tall as the longleaf pine forests of the lower coastal plain in the southeastern United States (Harris 1980). Every coniferous genus occurring in these forests is represented by the largest and the longest-lived species of that genus (Waring and Franklin 1979). Because of the size and longevity of the trees, total above-ground biomass values exceeded 475 tons per acre (1070 tons/ha) (Grier and Logan 1977). The key to the large size and biomass values is not an exceptional annual growth rate, but rather sustained height growth for over one hundred years. On higher elevation sites,

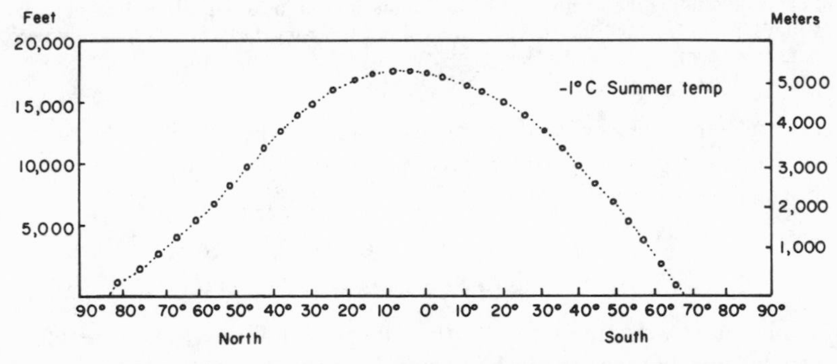

Figure 3.3 Twenty-eight degree F (−1°C) average July temperature isotherm over the northern Pacific and average January temperature over the southern Pacific illustrating the relation between latitude and elevation (from Swan 1968).

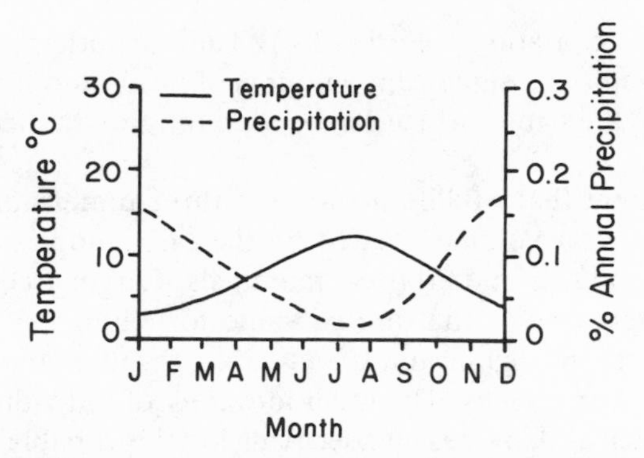

Figure 3.4 Asynchronous temperature and precipitation trends for the western Cascades demonstrating the nature of the Mediterranean climatic pattern (after Waring and Franklin 1979).

Douglas fir may exhibit substantial height growth into the second and third centuries of life (Franklin and Waring 1980).

Waring and Franklin (1979) have hypothesized that the large size and longevity of the trees may be adaptive strategies to contend with a climate where peak radiation occurs six months out of phase

Figure 3.5 Relative height of mature forest canopies with Douglas fir–western hemlock on the left, mature longleaf pine in the center, and a slash pine plantation on the right.

with peak precipitation (see fig. 3.4). They hypothesize that the large trees can store significant amounts of intra- and intercellular water during the winter which can be used for growth the following summer.

To the extent that wildlife habitat is a three-dimensional space, the larger habitat volume created by the tall canopy is of direct importance to birds and arboreal mammals. Larger trees can also contain larger cavities that are of value to mammals. The large fallen logs create significant travel lanes through thick second growth and over ravines. The great longevity of individual habitat structures such as den trees and decaying logs is probably an important stabilizing factor to individuals and family lineages of wildlife.

Conifer Dominance

Because of the latitude, altitude, high volumes of snow, and natural lightning-fire frequency, it is not surprising that the forest should be dominated by conifers. But because of drought conditions during the summer, conifers have assumed a dominance over hardwoods that is unrivaled elsewhere in the world (Franklin 1979; Franklin and Waring 1980). Despite an overall low tree species diversity, the number of conifer species, twenty-five, is several times that in any other region of North America. These species have radiated into nearly every environmental niche (fig. 3.6). This diversity, coupled with the low angiosperm diversity characteristic of the latitude (Rejmanek 1976), means that the conifers outnumber the hardwoods a thousand to one (Kuchler 1946 [in Franklin 1979]) and contain a thousand times more biomass (Waring and Franklin 1979). Since conifer dominance is particularly great on mesic sites, total species richness and species diversity are low on these sites. Plant species diversity appears to be greatest on more xeric and more hydric sites (Zobel et al. 1976).

The twelve hardwood tree species that occur in the region are generally limited to riparian strips where water is available during the summer growing season or other sites where they hold a competitive advantage. Their abundance decreases with ascending elevation and may well influence the reduction of vertebrates with increasing elevation. Since hardwoods are generally believed to support more species and a higher abundance of wildlife than conifers (Thomas et al., 1975; Harris 1980), the extreme dominance of conifers has direct implications for wildlife.

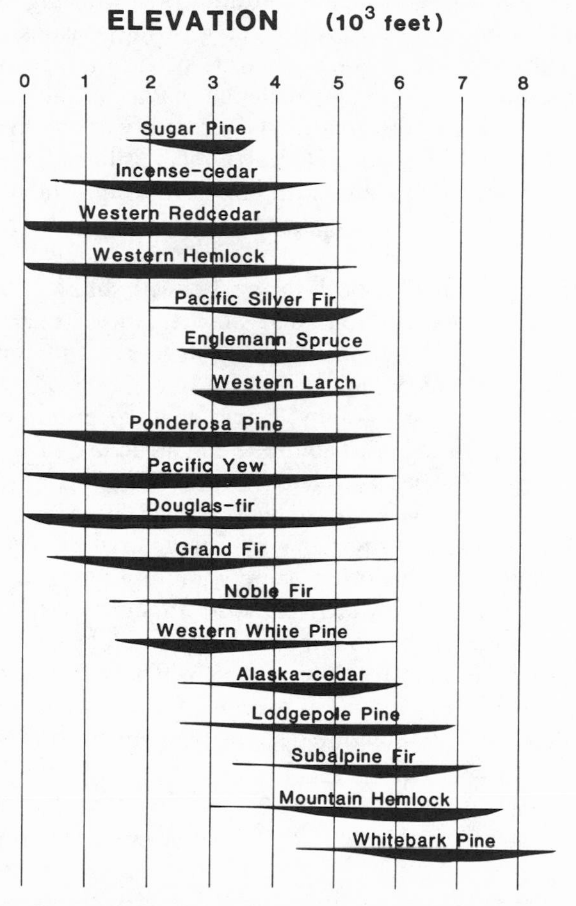

Figure 3.6 Ordination of principal conifer species along the elevational gradient at approximately 44° N latitude in the western Cascades (from Langille et al. 1903).

Highly Dissected Topography

Since drainage density is a strong function of relief, and the Cascades are geologically young and steep, the landscape is highly dissected by streams and stream channels. Because the main range lies in a north-south direction, the principal streams and secondary ridges lie in an east-west plane. This creates a major dichotomy between north-facing and south-facing slopes, which accentuates the importance of latitude and makes aspect doubly important. Whereas north-facing slopes are generally wetter and cooler, with snow pack occurring earlier in autumn, south-facing slopes are much drier, hotter, and have longer growing seasons (fig. 3.7). These factors greatly affect regeneration, forest succession, and stand structure. Tesch (1975) concluded that aspect affected harshness of site, which in turn governed the rapidity of site invasion and thus the stand age structure. On north-facing slopes, natural stand regeneration occurs rapidly with high stocking, and thus these stands tend to be dense and approximately even-aged. The amount of understory is negligible in the denser of these stands. East-facing, west-facing, and finally south-facing slopes exhibit progressively longer establishment periods and depart more from even-aged conditions. A highly variable canopy structure and composition with a more dense and varied understory may then result. Processes such as these allow topography, slope, and aspect to greatly alter stand composition, age distribution, and structure of stands even within the same species as well as between different species. Since aspect affects snow pack and thaw rates that affect runoff, it also determines stream and riparian strip characteristics. Thus the discreteness and juxtaposition of one community type (conifer) with another (riparian hardwoods) are also affected by the dissected landscape. On level terrain, an acre is delimited by a quadrant 208 feet square. A land surface sloping at 35° will have nearly 25% more area than the same horizontally measured unit if aerial mapping is used. Quantitative population or density measures of wildlife may also be affected by this.

In attempting to categorize and classify the different components of plant diversity in the nearby Siskiyou Mountains, Whittaker

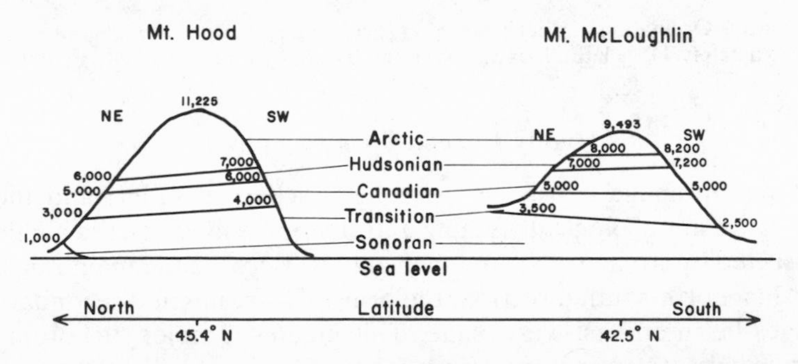

Figure 3.7 Comparative elevation of life zones illustrating the significance of latitude, elevation, and aspect in determining the nature of biotic communities of the western Cascades (adapted from Bailey 1936).

(1960) distinguished between the species richness of a particular stand or community and the species richness that results from gradients or patterns of environment. He labeled these "alpha" diversity and "beta" diversity respectively. Since contemporary forest management practices tend to reduce within-stand diversity while accentuating between-stand diversity, this concept holds major implications for forest management and animal ecology. Only the within-stand structural features of particular importance to vertebrates are discussed here. It must be borne in mind, however, that the combination of high latitude and dissected landscape amplifies the importance of aspect and creates the plethora of site-specific conditions leading to very high regional or landscape diversity (gamma diversity).

Structural Characteristics

Edaphic, climatic, and biotic factors are differentially important in determining characteristics of the vertebrate community, depending on elevation, access by humans, and other factors. For example, climate appears more important at higher elevations, whereas biotic interactions and habitat complexity are more influential at lower elevations. Only the biotic characteristics of special significance to wildlife are given here.

Uneven age and/or size distribution of dominants and subdominants create a rough canopy surface that may facilitate the interchange of gases, increase the amount of dew drip (and resulting energy in the form of heat of condensation), and enhance the interception of rain and snow (Fritschen and Doraiswamy 1973; Azevedo and Morgan 1974; Miller 1978). Along with the solar radiation that enters the mid- and ground-story levels, these factors allow vegetation growth throughout the vertical profile. Resulting herbs and shrubs on the forest floor are of direct importance to herbivorous wildlife. By supporting phytophagous insects, this vegetation constitutes an indirect support base for the many insectivorous vertebrates. This support is especially important during spring and early summer, when virtually all forest birds are insectivorous or carnivorous. Since hardwoods rarely occur in the canopy, the abundance of seed- and fruit-producing subdominants of value to granivorous wildlife species is inversely related to canopy closure (Anderson et al. 1969). In addition, vertical height distribution of foliage is widely accepted as a predictor of bird species diversity;

thus the enhanced structural complexity due to broken canopies probably leads directly to greater bird abundance.

Standing dead trees are of wildlife habitat value for several reasons (Thomas et al. 1979a). They create an opening in the canopy that allows light penetration to the ground. This is believed to be important for the reasons given above. They provide a deadwood medium for arthropods (especially burrowing beetles) that enhances the food base for insectivorous birds, provide essential perching sites for raptors and other birds, and are important for cavity nesters, which total about 25% of the western Cascade breeding bird species. Dead snags are especially important in the Northwest since no vertebrate species excavates cavities in live wood. All seven species of woodpecker resident in the Cascades excavate holes only in dead wood. Because of their "self-pruning" adaptations, conifers generally do not form cavities at the site of branch excision; thus few natural mechanisms exist for the creation of cavities in young Northwestern conifer forests. Cavities of nonanimal origin generally result from decay that enters the tree at sites of injury from lightning, fire, or abrasion from falling trees. All of these injuries tend to increase with stand age.

Standing broken-top trees, not necessarily dead, also contribute to wildlife habitat value. Because of their large size, most owls cannot use cavities created by woodpeckers and thus depend on the larger cavities associated with broken tops. The loss of leader dominance and resulting upturned branches are of value in suspending snow in the canopy and reducing the depth at ground level. This greatly enhances the habitat value of old growth during winter. These same upturned branches support the nests of the arboreal red tree vole.

Extensive surveys of decay and related damage in Douglar fir forests revealed a "decided increase in the amount of decay up to and including the 301 to 350-year age class. . . . For site II, with maximum rot, the most rapid increase in decay amounts to 23.1 percent of the gross volume in board feet between 300 and 350 years, or 0.46 percent a year. And for site II with average rot it is only 0.29 percent a year between 400 and 450 years" (Boyce 1932). Boyce (1932) further estimated that on site II lands, decay would equal growth increment at about 300 years of age. Stands under 100 years old usually do not contain significant amounts of heart rot and are thus of less value as potential cavity trees.

Whereas standing dead trees that contain cavities and heart rot are of great importance to the avifauna and some mammals, tree-base cavities are important to nonclimbing mammals. These cavities tend to be more abundant in old-growth forests and forests growing on wet sites. Butt-rot decay not only creates a den while the tree is standing but makes the tree more susceptible to wind throw, thus increasing the number of fallen logs.

Fallen trees and logs reach densities as high as 283 tons per acre (600 t/ha) (Edmonds 1979; Franklin et al. 1981; Grier and Logan 1977) in old-growth stands. Just as greater foliage height diversity leads to a greater diversity of birds and perhaps arboreal vertebrates, a heterogeneous ground layer creates better habitat for cursorial and fossorial vertebrates. Down logs provide critical habitat for certain amphibian and reptile species and serve other important functions for birds and mammals. Fallen tree boles serve such purposes as travel routes for mammals (Olszewski 1968), drumming and display sites for ruffed grouse, a concentrated source of invertebrates and small vertebrates for carnivores such as bears (Maser et al. 1979), and nest protection for ground-nesting birds (Bowman and Harris 1980).

Understory and epiphytic vegetation is an essential component of wildlife habitat, especially in the Douglas fir forests where angiosperm diversity is so low. Herbivores are directly tied to the presence of herbage, and the greater the interspersion of food resources with cover resources, the higher the habitat value. Angiospermous plants produce all of the nectar that is consumed by nectivores (e.g., hummingbirds) and the majority of fruits and seeds (mast) for granivores and omnivores.

The small number of hardwood tree species that occur in the Douglas fir region has direct implications for wildlife, as will be illustrated by the oaks. A comprehensive review of wildlife food habits (Martin et al. 1951) demonstrated that *Quercus* was the single most important genus of North American wildlife food plants. Only one species, Gary oak, occurs in Washington and throughout most of Oregon. This may be contrasted with other coniferous forest regions, where numerous species occur. For example, twenty-five species and four subspecies of oak occur in the state of Florida. Because of this low hardwood diversity, understory angiosperms are all the more important as producers of seeds and fruits.

Epiphytic foliose lichens and fungi are principal dietary items of the northern flying squirrel (Maser et al. 1978b). Upon falling to the ground, the lichens are important winter food for Columbian black-tailed deer, Roosevelt elk, and other mammals (Rochelle 1980; Schoen et al. 1981). Foliose lichens such as *Lobaria oregana* are particularly important because of their nitrogen-fixing ability and thus higher protein content (Denison 1973). A single 400-year-old Douglas fir tree may support thirty pounds (13.5 kg) of lichens and ten pounds (4.5 kg) of bryophytes (Pike et al. 1977).

Mycorrhizal fungi are a symbiont of major importance to most higher plants, acting to increase nutrient uptake efficiencies and productivity of forests. The growth rate of inoculated trees is usually much greater than that of noninoculated trees. Epigeous fungi (growing above ground) are the most common types in many forest ecosystems. Their spores are typically wind-dispersed. Hypogeous fungi (truffles) grow beneath the soil surface, and it seems that this group is most important in the Douglas fir region (Trappe and Fogel 1978).

The hypogeous fungi have evolved a considerable dependence on animals for spore dispersal (Maser et al. 1978a). The different aromas associated with truffle species are believed to serve as attractants to mammals that consume them and in turn disperse the spores, which are conveniently located in piles of dung (Trappe and Maser 1977). Many species of mammals are involved in the trafficking of truffles, but species such as the California red-backed vole only occur in coniferous forests of this region and depend almost totally upon hypogeous fungi for food. This species declines rapidly after clearcutting and only builds significant populations after a coniferous forest is reestablished (Maser et al. 1978b). Although the overall dependence of hypogeous fungi on mammals is unknown, the role of mammals in dispersing fungal spores is probably considerable.

The Unique Combination of Characteristics

While any one of the above-mentioned characteristics might distinguish the old-growth Douglas fir forest from other forests in North America, it would not necessarily distinguish a natural old-growth Douglas fir ecosystem from a young managed stand. It is possible to have tall massive trees and high broken canopies in managed stands. It is possible to achieve conifer dominance with hardwood species limited to riparian strips. Perhaps it is possible to have uneven age distributions, standing snags, broken top trees, down logs, understory vegetation, epiphytic lichens, and butt-rot cavities in young managed stands (table 3.2). Thus, when considered individually, none of the above characteristics is unique to old-growth forests. However, the old-growth ecosystem is unique in possessing all these attributes simultaneously, and in addition no doubt has many others that we are unaware of.

No matter how successful one might be in creating wildlife habitat for one or even 200 wildlife species, a planted, managed stand may never take on all the characteristics of an old-growth ecosystem. The distinction between old growth and old age must be kept in mind. Whereas old growth refers to an ecosystem type with numerous structural and functional attributes (both known and unknown), old age refers simply to a chronological state that may or may not have a direct relation to ecosystem attributes. To identify old-growth habitat characteristics that are important to wildlife is a meaningful and worthwhile endeavor. It should not, however, be confused with the arguments for conserving old-growth ecosystems.

In addition to the characteristics of the old-growth system listed above, there is yet another of importance to wide-ranging wildlife species. The Douglas fir forest remains the most expansive tract of interconnected, unlogged forest in the conterminous United States. It still spans an area 500 miles (800 km) by 100+ miles (160+ km), and it still occupies approximately 25% of the National Forest acreage in this region. The large expanse, inaccessibility, and partial connectivity are perhaps its most important characteristics. They are also the ones most in jeopardy.

Table 3.2 is found on p. 24.

TABLE 3.2

Distinctive wildlife habitat characteristics of natural old-growth Douglas fir forests of western Oregon and western Washington (see also Franklin et al. 1981).

- Combination of high latitude and rugged terrain, accentuating the importance of aspect and elevation in creating a highly diverse landscape mosaic of habitat types.

- Conifer dominance with hardwood tree species mostly limited to lower elevations, wet sites, and/or riparian strips. Because little rainfall occurs during the summer breeding season, the value of these hardwood riparian strips located near water is greatly enhanced.

- Long-lived, massive conifer species that create large structural components of habitat and a twofold to threefold greater habitat volume per unit of land area. The longevity and massivity of trees, snags, logs, cavities, etc., contribute time stability to the habitat.

- Broken canopy with uneven age distribution of dominants and subdominants that creates a vertical foliage distribution of primary importance to birds and a patchy understory of value to all vertebrates.

- Numerous snags that provide roosting and nesting sites for birds and a medium for wood-burrowing insects that provide the food base for many insectivorous species. A high incidence of decay in live trees is important for the same reasons.

- Broken top trees that provide nesting cavities for large bird species such as the spotted owl. The upturned branches resulting from loss of leader dominance create nesting habitat for mammals such as the red tree vole and suspend snow in the canopy thus reducing snow depth on the ground.

- Abundant foliose, epiphytic lichens that constitute the food base for certain mammals such as the northern flying squirrel and provide critical winter forage for species such as black-tailed deer.

- Hypogeous mycorrhizal fungi that constitute an important food source for several mammal species and the predominant food of the California red-backed vole.

- Fallen logs that constitute an important medium for arthropods, which serve as a food source for many vertebrates. Logs serve as habitat for vertebrates that live in the decaying wood and also perform other ecological functions.

4

Forest Trends and Patterns

The first water-driven sawmill of the western Cascades region was apparently constructed in 1827, with the first export shipments of wood dating from 1830 (Kirkland and Brandstrom 1936). Puget Sound's first water-driven sawmill was built in 1851 and was followed by mills at Seattle and Coos Bay in 1853 (Sargent 1884). The steam-driven sawmill was operative shortly thereafter, and the lumber industry boomed along with the California gold rush (Twight 1973). Mass-produced Douglas-fir wood products were exported all over the world at this time (Kirkland and Brandstrom 1936). Still, by the turn of the century only 1.6% of the Washington forest and an even smaller percentage of Oregon forest had been logged (Plummer 1902). As of 1928, about 4 million acres (16,000 km^2) of original old-growth had been logged (Fery 1928), and clearcutting was progressing at a rate of 200,000 acres (810 km^2) per year. About 37% of the original 19 million acres (77,000 km^2) had been logged by 1936 (Kirkland and Brandstrom 1936). By 1946, about 66% of the original forest acreage was cutover and "splendid old-growth timber remained on about 3.5 million acres" (Kirkland 1946).

Ownership

Over 90% of early logging occurred on private lands (table 4.1), and by 1928 the logging pattern was cast. The highest volume and most accessible timber was privately owned. When coupled with the concept of holding cost, this compelled owners to cut the superior low-elevation sites and leave the lower-quality wood. Gannett (1902, 8) reported that "the Washington lumberman is skimming the cream. He is taking only the best and heaviest of the timber and that which is most accessible. . . ." Already by 1928 it was observed that the "ultimate outcome in this region will prob-

TABLE 4.1
Volume (million bd. ft./yr.) of Douglas fir sawtimber cut from different forest ownerships in the western Cascades during the years 1925–33. Data from 1933 forest survey, Andrews and Cowlin 1940.

Source	Volume	% of cut
Private land	6,592	89
National forest	225	3
OR and CA (O & C) lands	245	3
Washington State land	175	2
Indian land	160	2

ably be to continue to log the lower elevations clean but there will no doubt be a tendency to use the selection system . . . in rougher country" (Fery 1928, 22). By 1936, "growing stock has now been removed from . . . the most accessible and, for the most part, the highest quality timberlands in the Douglas fir region. The result is that the operable timber zone has been pushed back into generally rough areas, remote from the manufacturing centers and principal shipping outlets" (Kirkland and Brandstrom 1936, 119).

Foresters of the day used the terms "large old-growth" and "small old-growth" to distinguish between the volumes on different sites and different capability classes. "A very large proportion of the original area of large old-growth was privately owned . . . most of the small old-growth is on national-forest land, little of which has been cut over . . . although it produces some clear lumber, this class as a rule is much inferior in quality to large old-growth" (Andrews and Cowlin 1940, 2). To elaborate further, the vegetation type mapping system used by forest management agencies relies on a series of symbols to denote specific characteristics. In the code D5≡ (Dee five three bar) the D denotes Douglas fir, 5 denotes a full stocking level, and ≡ denotes the largest size class of trees currently existing. It is of interest to note that earlier in this century the "large old-growth" was typed as D5𝌆. The five bars indicated the largest size class; this code symbol has since been dropped from usage, presumably because of too few opportunities for its use.

The degree of accessibility was measured by the economics of extraction. Timber that was economically feasible to extract prior to 1925 was placed in class I, that which could be logged at a modest loss at 1925 prices was assigned to class II, and all other timber was class III. As of 1933, between 60% and 70% of class I timber was

TABLE 4.2
Percentage of 1940 volume in the three economic availability classes and two ownership classes. Data from 1933 forest survey (Andrews and Cowlin 1940). Class I = economical to log prior to 1925; class II = could be logged at 1925–29 prices at loss of $5/mbf; class III = all other timber.

	Douglas fir		Pulp species		All other spp.	
Class	N.F.[a]	Other	N.F.	Other	N.F.	Other
I	38	62	30	62	37	70
II	52	34	54	34	52	26
III	10	4	16	4	11	1

a. N.F. = National Forest.

privately owned while about 35% was in public ownership. Higher percentages of class III land occurred in the national forests (table 4.2). At a slightly later period the same cutting pattern prevailed on public lands. The first sweep of cutting occurred in valley bottoms, the second sweep higher up the slope, and the third sweep still higher up the mountainsides (fig. 4.1).

Depletion

The issue of depletion was not new to the 1970s. At the time of the first official forest survey in 1933, it was reported that the "saw-timber volume of the Douglas-fir region is being depleted about four times as fast as it is being replaced by growth" (Andrews and Cowlin 1940, 43). "Ample evidence exists that extensive clear cutting as at present practiced will not accomplish this, but on the contrary it will result in depletion of the resource and loss of most of the capital values dependent thereon" (Kirkland and Brandstrom 1936, 121).

For the last thirty years, annual loss and removal of Douglas fir sawtimber from western Washington and western Oregon has averaged three times annual growth. During the 1950s, private industry made large gains toward balancing the cut and the growth because early regeneration stands were highly productive. Small gains were made on public lands in the 1960s, but overall the trend has been toward greater deficit cutting (fig. 4.2). While there has been a modest 5% reduction in total commercial forest acres, the reduction in net volume of softwood growing stock has been 18%, the reduction in volume of softwood sawtimber has been 21%, and the

Figure 4.1 Photographs of Iron Mountain drainage in the Gifford Pinchot National Forest, Washington, illustrating the sequential nature of cutting in valley bottoms first and working progressively higher up the slopes (photos courtesy of U.S. Forest Service, Region 6 office).

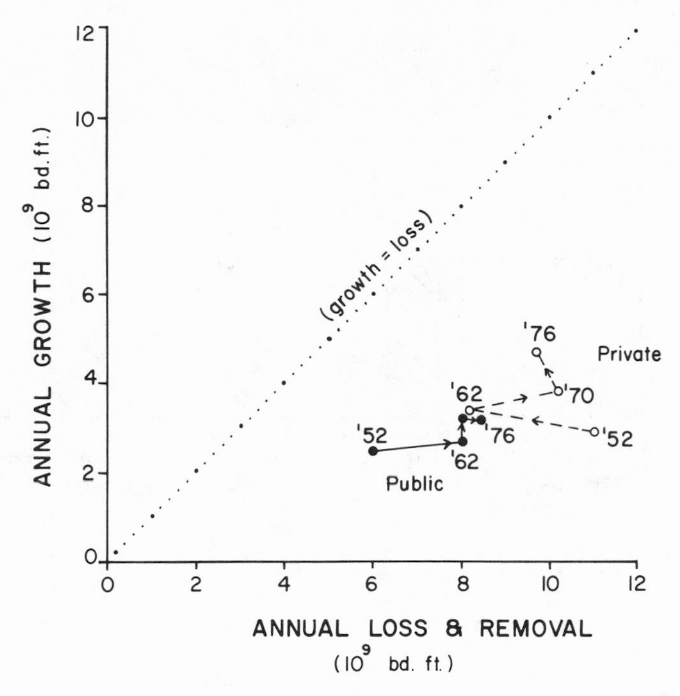

Figure 4.2 Disparity between annual growth, and annual loss and removal of softwood from private and public lands in western Washington and western Oregon. (Data from tables 34, 35, 36, U.S.D.A. 1978; figure from Harris et al. 1982.)

reduction in large-diameter-class softwood has been 34% (fig. 4.3). This trend may continue in the near future, as "public old-growth harvest substitutes for private young-growth harvest over the next 25 years" (Adams 1977; Brodie et al. 1978). A different analysis suggests that "only in the north coast timbershed in western Oregon and in the three eastern Oregon timbersheds could harvesting continue at the current level for the next 30 years . . . for western Oregon as a whole, this projection indicates a decline of 22 percent by the year 2000" (Beuter et al. 1976). Only when the recently planted stands reach high levels of mean annual increment will the current annual cut be matched by annual production.

Current Distribution

At the turn of the century, about 90% of western Oregon and perhaps 85% of western Washington forests were considered old

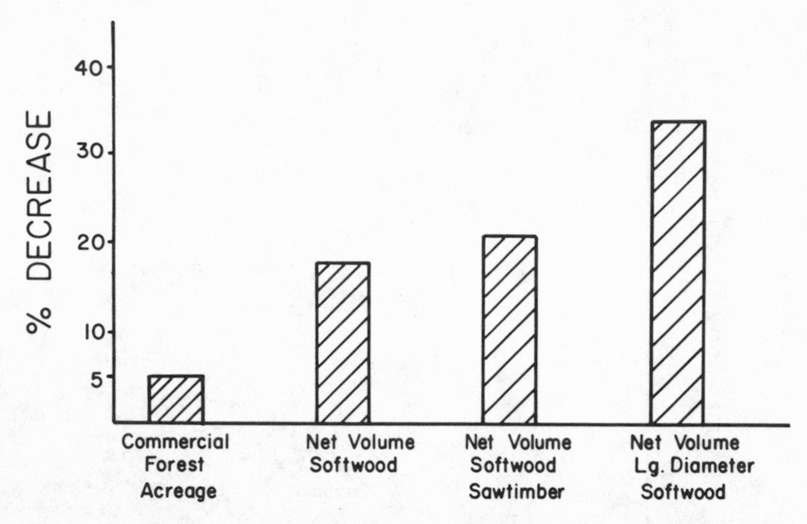

Figure 4.3 Relative decrease in softwood products occurring on commercial forest lands of western Oregon and western Washington from 1952 to 1977. Sawtimber volume is defined as the saw log portion of live trees containing at least one twelve-foot saw log or two noncontiguous saw logs each eight feet or longer. Softwood sawtimber trees must be at least nine inches (13 cm) dbh (U.S.D.A. 1978).

growth (Gannett 1902; Plummer 1902; Langille et al. 1903). The percentage of trees in the smaller age classes was very low. Because of cutting, the number and volume of old trees have obviously decreased. Because of severe regeneration problems during the early decades of this century, there exists a much smaller proportion of acreage in the thirty- to eighty-year age class than one would expect in an equilibrium regulated forest. Intensive replanting during the last thirty years coupled with heavy cutting of old growth has increased the acreage and number of trees in the young age class to a level much higher than would occur in the equilibrium regulated forest. Most recent forest inventory statistics for western Oregon and western Washington reveal that less than 1% of softwood trees occurs in each of the six size classes greater than thirteen inches (33 cm) dbh. In addition, approximately 85% of all trees are less than five inches (13 cm) dbh (USDA 1978, 29). Although normal mortality as the trees age necessitates that more trees occur in younger age classes than in older classes, this degree of skewness has significant implications for wildlife which will be discussed below.

Based on limited sampling, a recent analysis of stocking by age and ownership class confirms that private ownership, Washington

Department of Natural Resources, and Bureau of Land Management (BLM) lands have less acreage in the older age classes and more acreage in the middle age class than occurs on Forest Service lands (fig. 4.4). This disparity exists because intensive cutting on private land preceded intensive cutting on National Forest lands by several decades.

Analysis of acreage in the various stand age classes on BLM lands in western Oregon supports the same conclusion. The very limited acreage in the sixty- to one-hundred-year age class means continued harvesting pressure on the old-growth forest for the next several decades (fig. 4.5). Not only is the amount of old-growth habitat being rapidly reduced, but the percentage of acreage occurring in very young regeneration stands is probably as high as it will ever be. As the millions of acres replanted during the last three decades mature into the middle and later years of rotation, wildlife species that abound in regeneration stands will decrease. This suggests that game species such as black-tailed deer and Roosevelt elk may decline significantly along with the acreage of foraging habitat.

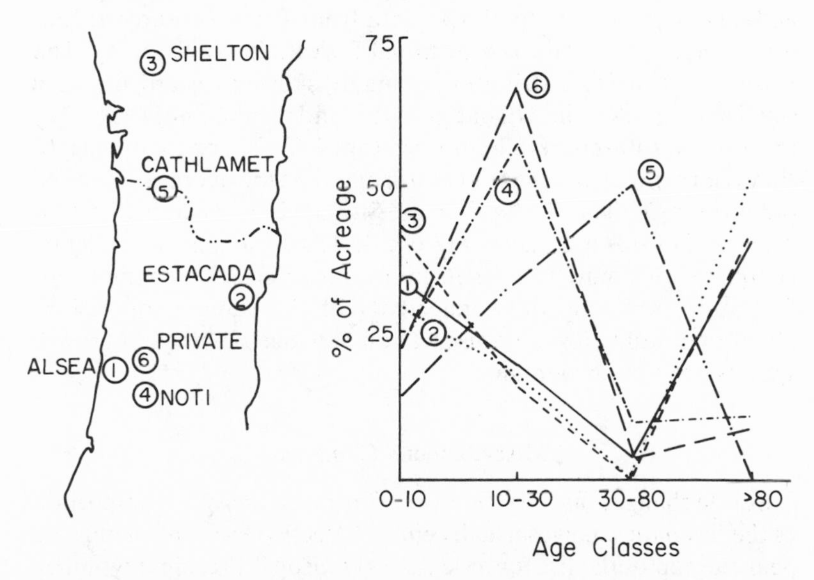

Figure 4.4 Distribution of softwood forest acreage by age class for six areas in the western Cascades. Site 1: Alsea Ranger District U.S.F.S.; site 2: Estacada Ranger District, U.S.F.S.; site 3: Shelton Ranger District, U.S.F.S.; site 4: Noti Resource Area, B.L.M.; site 5: Cathlamet District, Washington D.N.R.; site 6: private ownership. (Data from Lang 1980).

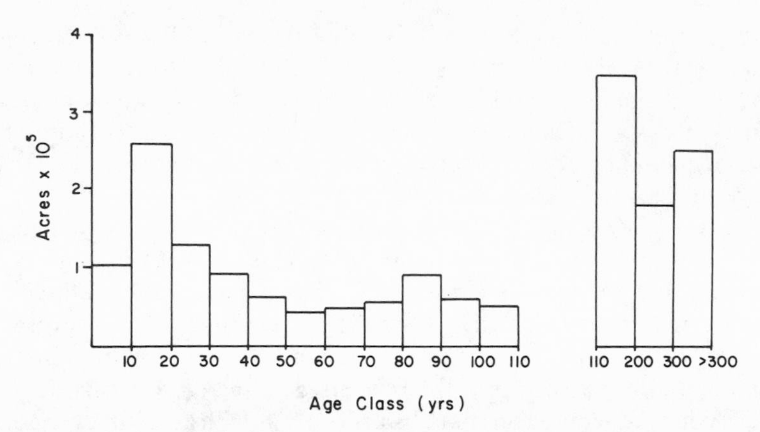

Figure 4.5 Bureau of Land Management forest acreage in western Oregon by stand age class (modified from Luman and Neitro 1980).

A somewhat different kind of analysis shows a result of old-growth cutting that can occur if mitigative actions are not taken. Old-growth Douglas fir (DO) acreage in the Siuslaw National Forest has been reduced to 3.3% of total forest acreage (Douglas fir and mixed old growth total 6%; data from Total Resource Inventory System [TRI], Siuslaw National Forest, Corvallis, OR). The existing patch size distribution is highly skewed toward the very small acreages. Of the 319 old-growth stands remaining, 196 (61%) are less than 40 acres (16 ha) in size (fig. 4.6). There are only eight stands larger than 350 acres (140 ha) and thus the average size of all old-growth Douglas fir stands in the Siuslaw is 68.2 acres (27.6 ha). The median size is 31 acres (12.6 ha). These small stands may be adequate for animals of small home range size, but without information regarding the connectivity of the stands, conclusions about their utility for the conservation of wider-ranging old-growth species are not encouraging.

Miscellaneous Changes

Subtle changes that are widespread in nature may be as important as the dramatic changes usually cited. Effective fire control must be near the top of the list for at least two reasons. Presumably, much less acreage is lost to catastrophic crown fires now than was formerly the case. Similarly, the reduction of nonlethal, low-intensity fires has no doubt altered the ecosystem structure and function of considerable forest acreage (Edmonds 1979).

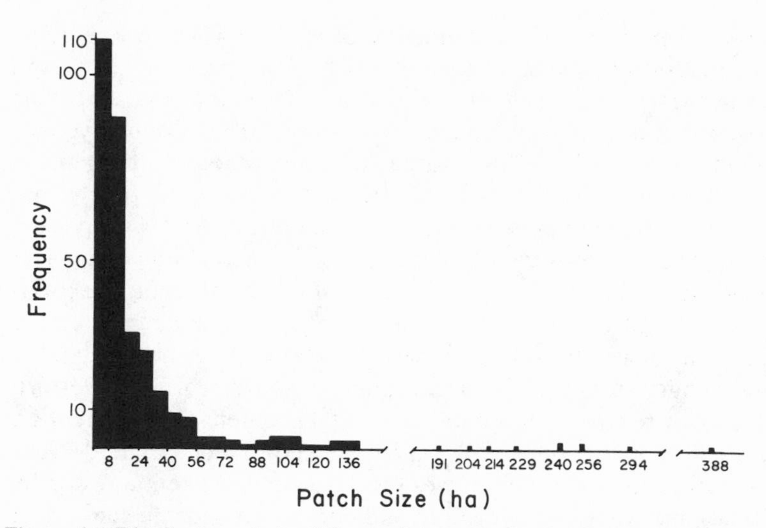

Figure 4.6 Distribution of the 319 remaining old-growth stands in the Siuslaw National Forest by size class. (Data for 1981 from Total Resource Inventory System, Siuslaw National Forest.)

Buechner (1953) described many additional changes that have occurred throughout the state of Washington in recent times. Although many of his points are not relevant to the western Cascades per se, the list is truly impressive. As many as 10 to 15% of the vegetation species growing without cultivation are of alien origin. "Hundreds of new kinds of insects . . . about 168 species of beetles" have been introduced (Buechner 1953, 172). According to Buechner, avifaunal changes involve (1) extirpation of certain large species from the area (e.g., California condor, whooping crane); (2) naturalization of three game birds (chukar Partridge, Hungarian partridge, ring-necked pheasant) and three nongame birds (English sparrow, rock dove or pigeon, and the European starling); (3) naturalization of four North American species not native to the region (scaled quail, bobwhite, valley quail, mountain quail); and (4) changed abundance of several species. Notable changes in the mammal fauna previous to 1951 involve (1) marked increases in populations of at least seventeen forms; (2) extirpation of several species (especially large carnivores); and (3) naturalization of perhaps five North American species not native to the area.

Throughout the first sixty years of this century the impact of forest practices on wildlife was either not apparent or was considered to be beneficial. There are at least four reasons for this: (1) old growth remained as a predominant forest type while regeneration and second-growth acreage was much less dominant; (2) large carnivores were considered a threat to human interests and their demise was not only sanctioned but encouraged by a bounty system; (3) the term "wildlife" was implicitly synonomous with "game," and therefore references to "wildlife" almost always implied game species; and (4) in part because of the emphasis on game species, old growth was considered to have little wildlife habitat value. The silviculturist for the Pacific Northwest Forest and Range Experiment Station observed that "large, continuous bodies of heavy timber are virtually biological deserts" (Isaac 1952, 14). In 1960, a preliminary survey of fish and wildlife resources of northwestern California reported on numerous habitat problems but no mention was made of the timber cutting issue (Anon. 1960). A mid-1960s habitat status evaluation in western Oregon listed ten noteworthy developments affecting wildlife and concluded that the sum effect of most logging had been to increase valuable wildlife habitat (Aney 1967). A similar evaluation concluded that "except for ducks and certain furbearers, game populations in the basin are believed to be the largest in history. . . . The beneficial effect of logging on deer habitat have [*sic*] greatly enhanced the recreational opportunities for the hunter" (Hutchison et al. 1966, 34).

Earliest appraisals that went counter to this concensus were signaled by Wight (1974), Meslow and Wight (1975), Forsman and others (1977), Meslow (1978), Luman and Neitro (1980), Meslow and others (1981), and Schoen and others (1981). Although the focus of their concern was on nongame species, the work of Schoen and his coauthors (1981) involved a game species.

At least in some circles, it is frequently asserted that clearcutting is analogous to the catastrophic fires that occurred naturally, and therefore contemporary forestry practices essentially mimic the natural situation. This analogy is incorrect for several reasons, but even if it were true it would be tangential to the issue. Early accounts report that fire had destroyed the timber on only about 10% of the forest acreage at any one time (Gannett 1902; Plummer 1902; Langille et al. 1903). This meant that the landscape was predominantly mature or old-growth forest. Currently, the proportions are reversed such that as little as 10% remains in mature or old-growth forest while 90% occurs as second growth. This ratio of old-growth and mature forest to young second growth is the parameter of most concern to wildlife biologists and ecologists.

Focus on the Willamette National Forest

Although not everyone would agree, the former senior forest economists of the region observed that "the national forests were originally created for the public domain and consist, generally speaking, of lands that up to the time of their withdrawal had not been considered desirable for private investment" (Andrews and Cowlin 1940, 38). Perhaps because of this, by 1933 half of the region's old-growth Douglas fir occurred in the upper Willamette Valley and in Douglas and Coos Counties (Greeley 1943). Focus on the Willamette National Forest will provide site-specific trends and allow site-specific evaluations.

The Cascade Range Forest Reserve was created in 1893, and as the largest of all the forest reserves it extended from the Columbia River to near the California border (Langille et al. 1903). In 1911 the Cascade Forest Reserve was subdivided into more manageable administrative units, of which two were the Santiam and Cascades National Forests. In 1933 the Santiam and Cascades Forests were combined to create the Willamette National Forest (Burns 1973). Containing approximately 1,600,000 acres (650,000 ha), the Willamette represents approximately 33% of the original Cascade Reserve.

The first recorded logging in the Willamette area occurred in 1875 on the middle fork of the Willamette River above Deception Creek. In 1905 about 14,000 mbf was logged from Gray's Creek for a stumpage price of 25¢/mbf (Anon. 1942). Logging increased dramatically in the 1920s but was greatly curtailed by the economic depression of the 1930s (fig. 4.7, appendix 2). Least-squares analysis of the annual cutting statistics from 1935 through 1980 reveals a geometric rate of increase in annual volume of harvest of 4.7% per year. This has resulted in a doubling of the volume cut every fifteen years (Harris et al. 1982).

At the turn of the century, the high-potential and high-volume sites occurred at lower elevations and in river valleys of the Willamette (figs. 3.2, 4.8). The spatial distribution of cutting has followed the general pattern described earlier. Cutting was initiated on the most accessible high-volume sites occurring in the river valleys and at lower elevations. Ninety percent of the cut during the first three decades of this century occurred below 4,000 feet (1,200 m) elevation. During the 1970s, 65% of the cut occurred above 4,000 feet elevation (fig. 4.9). Because of the lower volumes per

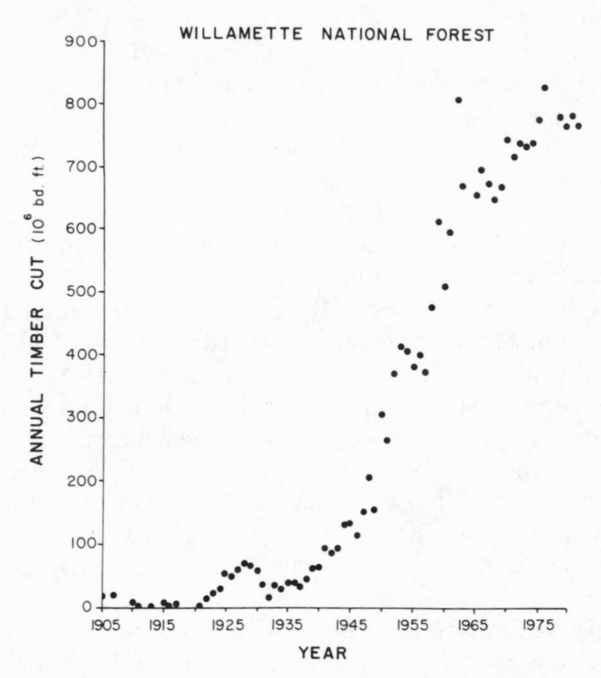

Figure 4.7 Annual timber cut in million board feet (2,358 m^3) from the Willamette National Forest. Data from historical cutting summary, timber management plan, Willamette National Forest (effective date FY 1977; Paulson and Leavengood 1977; figure from Harris et al. 1982).

Figure 4.8 (opposite) Old-growth and large second-growth distribution in the Willamette National Forest in 1936. Because the crest of the Cascades Mountain Range forms the eastern boundary, the gradient runs from lower elevations in the west to higher elevations in the east. Douglas fir is naturally limited to lower elevations (redrawn from Anon. 1936).

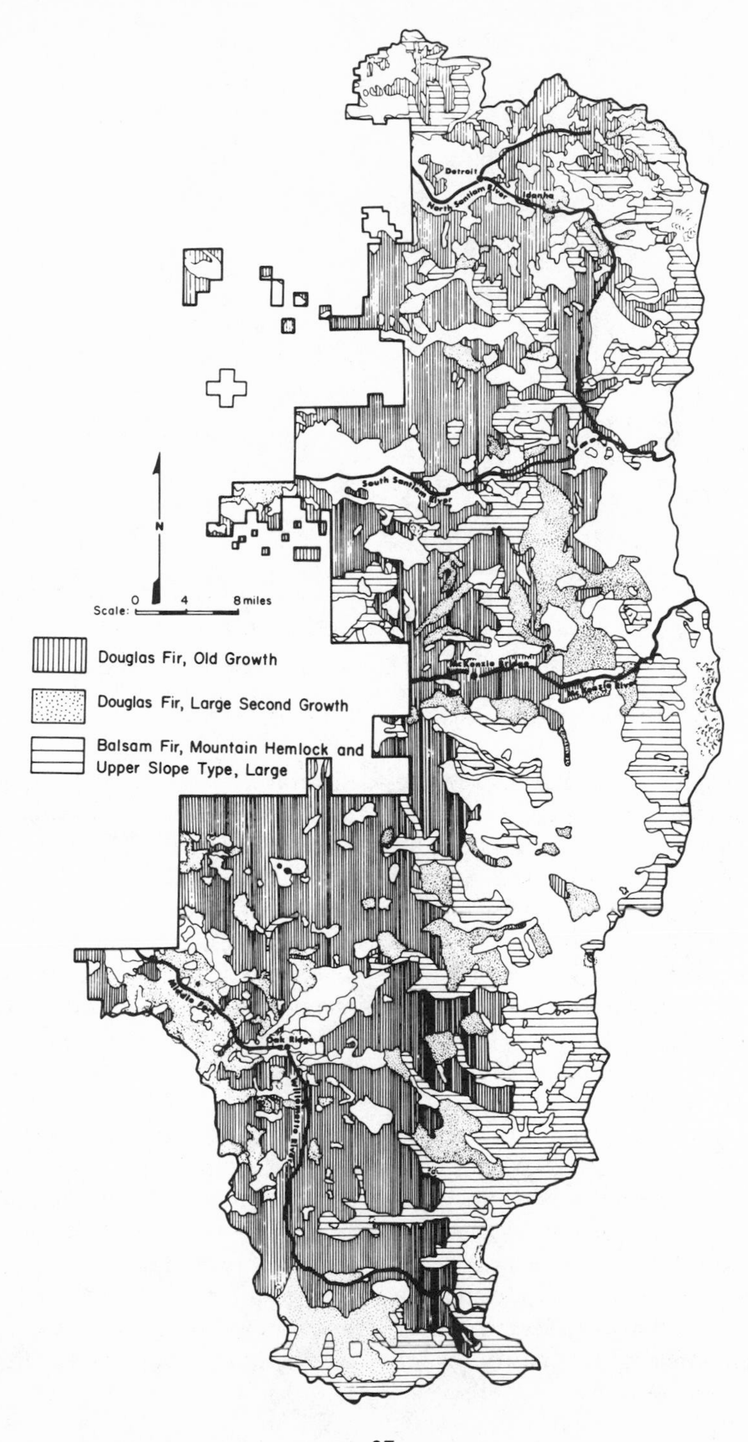
Detroit
North Santiam River
Idanha
South Santiam River
McKenzie Bridge
McKenzie River
Middle Fork
Oak Ridge
Willamette River
N
Scale: 0 4 8 miles
Douglas Fir, Old Growth
Douglas Fir, Large Second Growth
Balsam Fir, Mountain Hemlock and
Upper Slope Type, Large

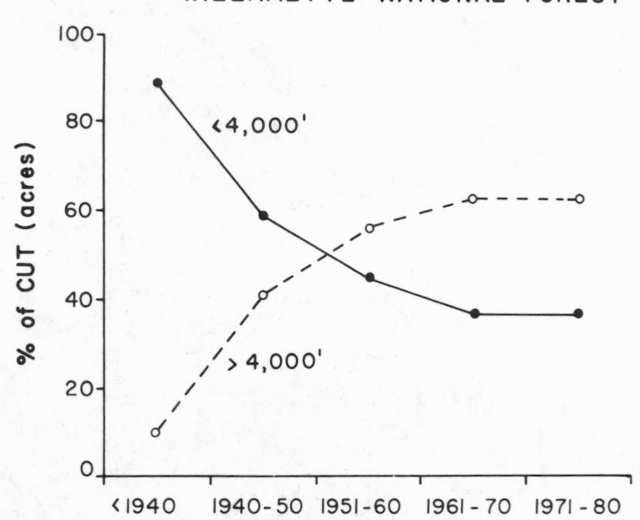

Figure 4.9 Percent of total annual cut from two elevation classes in the Willamette National Forest. Data extracted from Total Resource Inventory System, Forest Supervisor's office, Eugene, Oregon (figure from Harris et al. 1982).

acre now present, and because the average tree harvested is progressively younger and smaller (Tedder 1979), the increase in annual acreage cut has been five times greater than the increase in volume cut during the last forty years (appendix 3).

The pattern of cutting has three major consequences. Low elevation high-site potential old-growth has been reduced much out of proportion to its original occurrence. The total acreage of old growth has been greatly reduced. The nearly continuous nature of the original forest has been replaced by a patchwork of old growth with much-reduced continuity and connectivity between patches (fig. 4.10).

A sample of seventy-seven 3.9-mi^2 (10 km^2) quadrats taken from 1981 Willamette National Forest vegetation type maps (twenty of twenty-six quadrangle maps) suggests that about 26% of the Willamette remains in old growth (D5≡, DH5≡, TDH5≡, FD5≡). The distributional pattern runs the gamut from areas where old growth remains as the matrix and cut areas are infrequent, to the other extreme where clearcut areas, regeneration stands, and second growth are the matrix and old growth only occurs as isolated

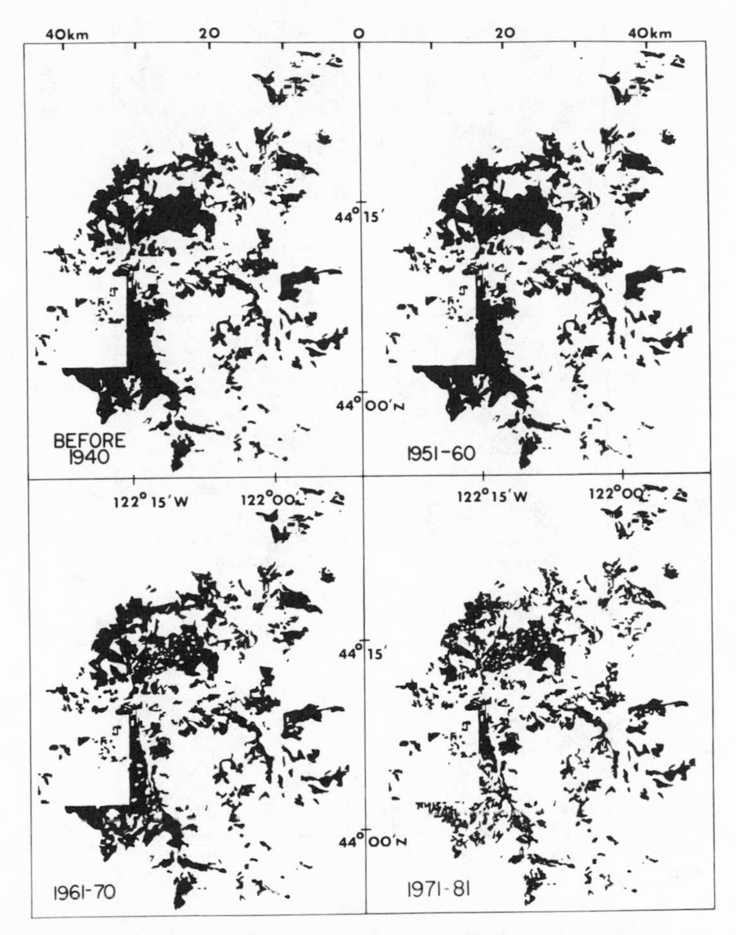

Figure 4.10 Distribution of old-growth stands on Blue River and McKenzie Ranger districts of the Willamette National Forest. (From aerial photos on file at Willamette National Forest, Eugene, Oregon.)

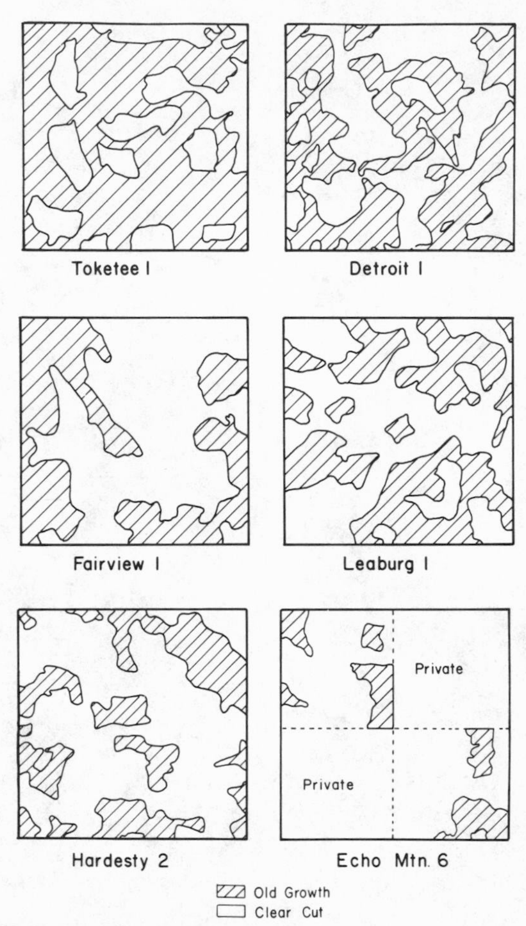

Figure 4.11 Representative samples of vegetation type maps for the Willamette National Forest. Each plot represents 3.86 square miles (10 km^2) with quadrat names referring to the quadrangle map from which the sample was drawn. The selection was chosen to portray the gradient of conditions from predominantly old-growth matrix to predominantly clearcut and regeneration stands (figure from Harris et al. 1982).

patches (fig. 4.11). Two of the 77 areas were more than 60% old-growth, 10 were more than 50%, 21 were more than 40%, and 35 were more than 30% old growth.

	no. plots	
0% <	21	< 10%
10 <	11	< 20
20 <	10	< 30
30 <	14	< 40
40 <	11	< 50
50 <	8	< 60
60 <	2	

The percentage of old growth ranged from 0 to 64 with the median and mean values both equaling 26%. If only "well stocked, large Douglas-fir" (D5≡) is considered, then the median percentage of acreage remaining falls to only 17. These data are consistent with regional statistics which suggest that 25% of the national forest acreage west of the Cascade summit supports stands greater than 250 years old with less than 10% of the timber removed (Sirmon 1982). Two representative aerial photographs are given as figures 4.12 and 4.13 and will serve as examples of the existing state.

Despite their inadequacies, these areas form an impressive arrangement of building blocks upon which a conservation strategy may be hinged (see fig. 6.14). If old-growth habitat islands are placed in strategic positions relative to these large preserves, the overall system will have a much higher chance of successfully conserving and enhancing the faunal resource. The integrated system of preserves and lower elevation old-growth islands must serve in place of the original "continent" of continuous boreal habitat and attendant species.

Figure 4.12 Aerial photo mosaic of forest conditions in the Blue River Ranger District of the Willamette National Forest, 1979. The apparent east-west line in the upper portion of the photo demarcates the boundary between private forest land above and national forest below. The middle section of the photo indicates about equal amounts of clearcut and old growth intermixed with little to no acreage in the mid-rotation-year classes. The Fawn Rock area in the lower portion of the photo demonstrates the nature of some relatively large expanses of remaining old growth. Unfortunately, the area has a high priority rating for removal.

Figure 4.13 Aerial photo mosaic of Blue River Ranger District in the vicinity of Hidden Lake (right center) and Cougar Reservoir (upper right corner), 1979. The sharp section line in the upper left (northwest) demarcates the boundary between private and national forest ownership. The uncut forest in the upper right corner lies within the Three Sisters Wilderness Area. Lower elevations (2,500 ft, 760 m) occur to the west while sites as high as 5,000 ft (1,500 m) occur in the southeastern portion of the photo area. The area appears to be highly cut over and is scheduled for additional high-priority harvest, but some potential for old growth and corridor planning still exists. Despite the occurrence of the south fork of the McKenzie and the major roadway along it, a corridor running from the French Pete area diagonally toward the center of the photo and then south into other areas of the Oak Ridge District can still be salvaged.

5

Animal Community Characteristics

Larry D. Harris and Chris Maser

The preceding chapters have described the unique and salient characteristics of the unmanaged forest, the structural components of old growth that enhance habitat value, the patterns of forest harvest, and present distribution. The following descriptions of animal communities, distributions, and habitat requirements only take on full meaning when viewed in the context of the earlier chapters. The southeastern coastal plain is geographically farthest from the northwest, numerous contemporary forestry analyses compare and contrast the northwest sawtimber forest with the southeastern pulpwood forest, and continental patterns of vertebrate distributions reveal a northwest-southeast trend line. For these reasons, occasional comparisons of the Douglas fir fauna with the southeastern coastal plain fauna will be made.

Nearly sixty years ago Shelford and Towler (1925) argued that a community approach was necessary for the organization and interpretation of the animal data available even at that time. The case is more compelling today, and this analysis will take such an approach. The objective is to describe both general and unique faunal characteristics and to demonstrate how patterns of distribution and abundance are tied to the topography and its vegetative veneer.

The Cascades region encompasses a great diversity of environmental conditions, ranging from marine to montane, xeric prairie to temperate rain forest, and congested cities to expansive wilderness, all within a distance of one hundred miles (160 km). Because of this environmental heterogeneity, the regional diversity of vertebrates is very high. In addition to the continued presence of North America's widest-ranging large carnivores, the region supports many unique localized specialists. The fauna of the Douglas fir forests is surely as distinctive as the forests themselves. This should come as no surprise, since animals do not stand apart from the forest but are an integral part of it.

General Characteristics

The Cascades region is about three times richer in mammalian species and about twice as rich in species of breeding birds as the coniferous forests of the southeastern coastal plain (fig. 5.1). The forty-eight families of breeding birds represent a greater richness at this level of classification than in any other area north of the Rio Grande (Cook 1969). This does not imply that more species occur within any single community type (Short 1979), but reflects a greater number of community types or at least greater diversity along the whole environmental gradient (Whittaker's gamma diversity). Because there are relatively few amphibian and reptile species compared to the high number of bird and mammal species, these taxa constitute a low proportion of the total vertebrate fauna. Again, to contrast the northwestern with the southeastern U.S., the

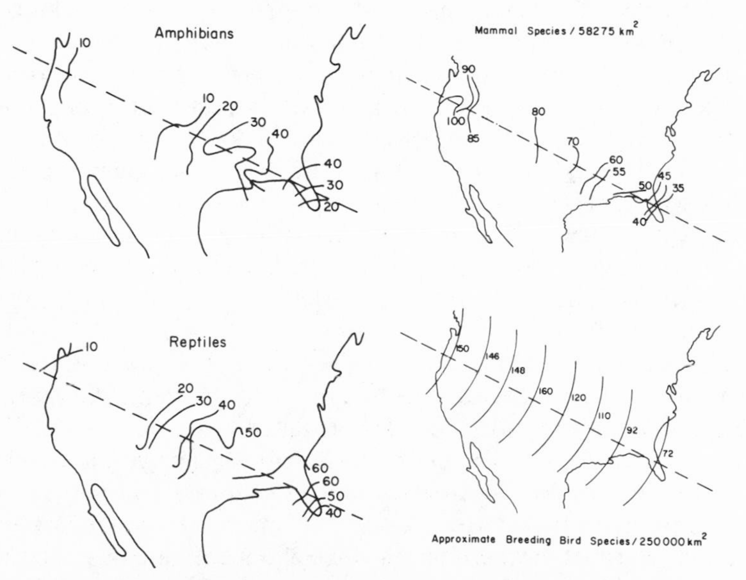

Figure 5.1 Comparative number of species of amphibians and reptiles and comparative species density of mammals and breeding birds along a gradient from northwestern to southeastern United States. Both coasts are dominated by coniferous forest. (Amphibian and reptile data from Kiester 1971; mammal data from Simpson 1964; bird data from MacArthur and Wilson 1967).

TABLE 5.1
Comparative number of species in various taxa occurring in Oregon and Florida. Source material from Robbins et al. (1966), Stevenson (1976), Behler and King (1979), Eltzroth and Ramsey (1979).

Taxon	Oregon	Florida
Amphibians	26	53
Reptiles	28	93
Permanent resident birds	172	105
Summer resident birds	88	39
Winter resident birds	48	98
Migrant birds	51	30
Breeding birds	251	156
Total mammals	125	63
Moles, shrews	13	4
Carnivores	20	13
Rodents	59	21

number of species in several taxa occurring in Oregon and Florida is given (table 5.1). Among mammals, the moles, shrews, rodents, and carnivores are particularly abundant in Oregon.

To allow discussion of specifics, the region of focus must be further restricted to Oregon, west of the Cascades ridge, because both the high Cascades and the Columbia River have served as major geographical barriers to fauna (fig. 5.2). Thus many species that occur in eastern Oregon (e.g., bighorn sheep, pronghorn) do not occur west of the Cascades, others occur in Oregon but not in Washington (e.g., nine species of mammals), while several that occur in Washington do not occur south of the Columbia River (Gordon 1966).

The western Cascades fauna derives from all major centers of geographic origin: species such as wolverine, marten, fisher, and pika derive from the Old World; short-tailed weasel, coyote, wolf, and red-backed vole are panboreal; chipmunks and flying squirrels are from boreal North America; and porcupines and opossums derive from South America. Among the mammals, about 75% of the species are of northern origin while about 25% are of southern origin (Gordon 1947; Yancey 1949). Among birds, as many as 80% of the species and over 80% of individuals inhabiting the coniferous forests are of boreal origin (table 5.2).

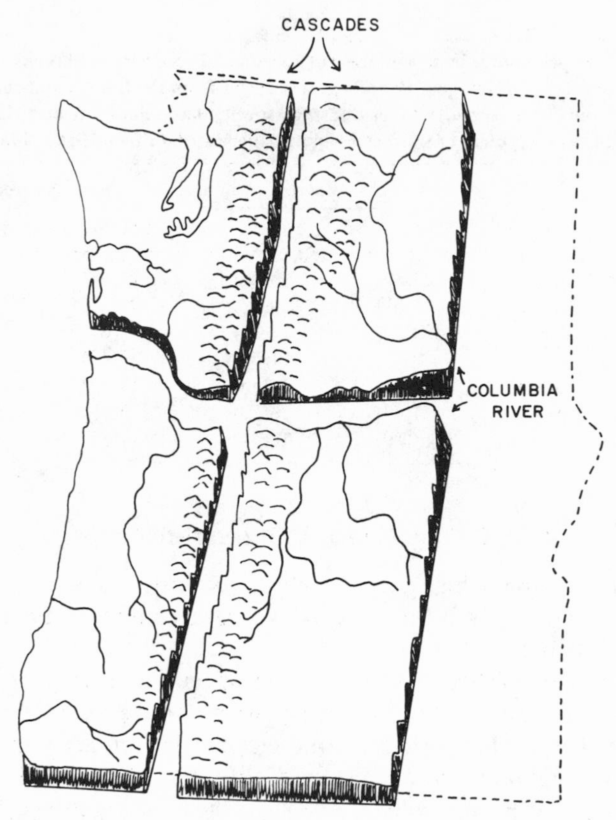

Figure 5.2 The Columbia River and the crest of the Cascades have served as biogeographic barriers dividing the Washington and Oregon biota into four quadrants.

Based on seven general vertebrate surveys relevant to the area (Yancey 1949; Sturges 1955; Hansen 1956; Beck 1962; Voth 1963; McKee et al. 1976; Airola and Barrett 1981), an average of 58% of the vertebrate species are birds, 30% mammals, 7% amphibians, and 5% reptiles (fig. 5.3). These studies span a wide range of conditions and community types and the numbers are averaged only to provide a reference point against which specific comparisons can be made. (Surveys from Steens Mountain and the Sierra Nevadas are not from western Oregon, but are included because of their presumed completeness.)

TABLE 5.2
Number (and percentage in parentheses) of species and occurrences of birds recorded in the Coast Range of Oregon. Boreal species derive from the tundra and coniferous forests of northern origin, whereas austral species have Sonoran and deciduous origins more typical of southern North America (data from Bratz 1950).

	No. species		No. occurrences	
	boreal	austral	boreal	austral
Noble fir	23(80)	6(20)	73(88)	10(12)
Douglas fir Assoc.	33(61)	20(37)	297(85)	52(14)
Riparian	22(56)	17(44)	76(67)	37(33)
Understory	9(45)	11(55)	46(43)	60(57)
Early regeneration	10(45)	12(55)	19(42)	26(58)
Late regeneration	24(48)	26(52)	73(38)	118(62)
Bracken fern	4(29)	10(71)	11(35)	20(65)

Unique Faunal Characteristics

In addition to the high species richness of birds and mammals, the old-growth Douglas fir forest exhibits several characteristics unique to North America.

The only two folivorous mammalian species that occur in North America are restricted to the forests west of the Cascades crest in Oregon. The world distribution of the genus *Arborimus* is limited to western Oregon and northwestern California (Maser et al. 1981). One of these species, the red tree vole, is the most highly specialized vole in the world and the most arboreal mammal in North America (Maser 1966; Johnson 1973; Maser et al. 1981). Individuals not only spend their entire lives in the canopy of Douglas fir trees but also many generations of individuals may inhabit the same canopy and nest site area during the life of a single tree, which may span several centuries. The species is primarily restricted to mature Douglas fir forests although it does occur in other tree species and in younger successional stages. The diet of the red tree vole is almost entirely restricted to Douglas fir needles (Maser et al. 1981).

The second species, the white-footed vole, is not as restricted to an arboreal way of life as is the red tree vole, and probably uses burrows in the ground (Maser et al. 1981). Nonetheless, recent analysis suggests that its diet is restricted to foliage of angiospermous plants (Voth et al. 1983). Riparian forests appear to be its preferred habitat and red alder leaves, for which the vole must climb, are its preferred food.

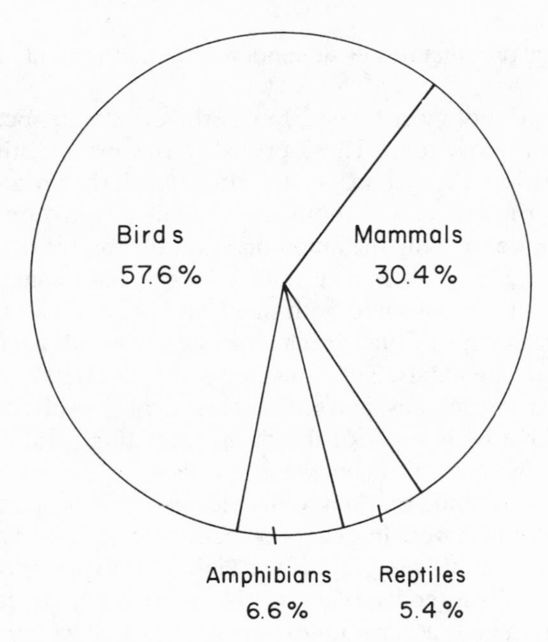

Figure 5.3 Relative percentage of species in each terrestrial vertebrate class based on seven faunal surveys in the Cascades region (see text for citations).

The only predominantly fungivorous species in North America occurs in the Douglas fir forests of western Oregon and northwestern California. The primary habitat of the California red-backed vole occurs in mature and old-growth forests. The vole spends most of its life in burrows beneath the forest floor where it obtains 90% of its diet: sporocarps (fruiting bodies) of below-ground fungi (Maser et al. 1978b). Many small mammals are known to consume fungi, but no species depends upon fungus to the extent that the California red-backed vole does. Several studies (Gashwiler 1959, 1970a, 1972; Goertz 1964; Hooven 1973) have reported a decrease or elimination of this species by clearcutting. This decrease is believed to result from disappearance of its food source, hypogeous fungi, after clearcutting (Maser et al. 1978b).

The only North American forest animal that consumes lichens predominantly also occurs in the western Cascades. Perhaps because there are no large-seed-bearing canopy tree species, the northern flying squirrel depends strongly on epiphytic lichens and fungi (McKeever 1960; Maser et al. 1978b; Maser et al. 1981), and this in large part dictates its dependence on mature and old-growth forests.

As is typical of most forests, few herbivore species occur in the Douglas fir ecosystem. Thus, probably the most distinguishing characteristic of the system's vertebrate fauna is the preponderance of carnivorous species without an obvious herbivorous support base. Considering only the amphibian, reptile, and mammal fauna of the H. J. Andrews Forest, 33% of the species consume plant material and 7% consume both plant and animal material, while 60% are carnivores. Considering all western Cascades amphibians, reptiles, and mammals, 28% consume plant material, 7% consume both plants and animals, and 65% are carnivores. Anderson (1970) reported that 69 to 87% of the birds inhabiting fir and hemlock forests of western Oregon were insectivores (carnivores) and Bowles (1963) found that ten of the eleven common species occurring in old growth were insectivores (carnivores); 71% of all species observed fell in this category. If the eighty-four species of coniferous forest birds listed by Wight (1974, figure 5.13) are considered along with the other vertebrates, then 65% of all western Oregon forest vertebrate species are carnivorous, 10% are omnivores, and 24% consume plant material. Thirteen (15%) of the avian species are large raptors (table 5.3) and only one of these (osprey) is not a permanent resident.

Twenty-two species of mammals with a weight greater than one kilogram are native to the Douglas fir region. Of these, fifteen are in the order Carnivora and at least twelve are true carnivores. The only herbivores are elk, black-tailed deer, porcupine, beaver, and muskrat. Compare this to the fauna of the northern Rocky Mountains, where the largest ten species consist of only one carnivore (grizzly bear), one omnivore (black bear), and eight herbivores. The twenty-two largest species of mammals in the northern Rockies include ten herbivores (elk, mule deer, white-tailed deer, pronghorn, moose, bighorn sheep, mountain goat, bison, porcupine, and beaver), two omnivores, and ten carnivores. Many of these herbivores are not forest-dwelling species but they do support forest-dwelling carnivores. No comparable array of herbivores has ever existed west of the Cascade crest in recent time. One might conclude that the fewer species of herbivores is compensated for by greater densities of individuals within species, but this is not the case.

TABLE 5.3
Thirteen species of raptors occurring in the Douglas fir forests of western Oregon (from Wight 1974) and approximate food habits based on literature review (Snyder and Wiley 1976).

	Diet			
Species	% bird	% mammal	% lower vertebrate	% invertebrate
Sharp-shinned hawk	93	2	1	4
Cooper's hawk	67	22	9	2
Goshawk	54	37	0	9
Bald eagle	19	4	75	2
Pygmy owl	13	23	2	61
Saw-whet owl	2	97	0	1
Long-eared owl	2	97	0	1
Great horned owl	6	78	2	15
Red-tailed hawk	8	50	4	37
Screech owl	3	66	1	31
Spotted owl[a]	4	38	1	57
Osprey	0	0	98	2
Merlin	25	0	0	74

a. Recent research by Eric Foresman indicates that this species consumes predominantly mammals.

The explanation for this large array of vertebrate carnivores in the western Cascades must rest on the diversity and magnitude of energy flow pathways other than herbivory. Three of these pathways have been identified above: (1) fungus production and mycophagy; (2) lichen production and its consumption; and (3) tree foliage consumption (fig. 5.4). Three additional pathways involve (1) nectar, fruits and seeds; (2) detritus; and (3) aquatic production coupled with thc annual migration of anadromous fish.

The fruits and seeds of the Douglas fir forest are not large and conspicuous. This may be why the concept of "mast production" is rarely mentioned in the northwestern wildlife literature and why most consumers of the mast crop are small birds or rodents. Gashwiler (1965, 1967, 1970b) monitored conifer seed production and consumption and found production to vary a hundredfold between years (0.02 lb. seed/acre [22 g/ha] 1957; 4.82 lb/acre [5.4 kg/ha] 1959). Seed-consumers seem to prefer Douglas fir seeds, and consume as much as 69% of the seed crop (average consumption =

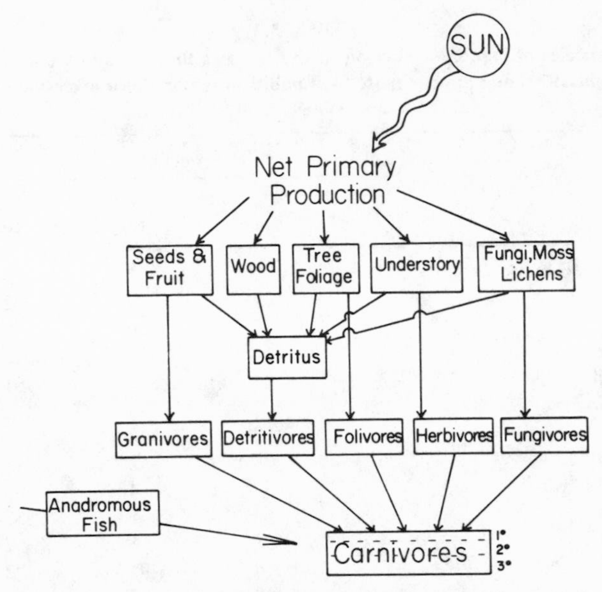

Figure 5.4 Principal energy flow pathways of the old-growth Douglas fir ecosystem that allow 65% of the vertebrate species to be carnivorous.

32,400 seeds/ha/yr). The principal seed-consumers were deer mice, several species of shrew, chickaree, Oregon junco, varied thrush, and five species of sparrow. The small mammal population density generally increased in years following abundant seed production (Gashwiler 1965).

Although no vertebrate feeds directly on detritus, many vertebrates depend on the detritivores inhabiting the forest litter and dead wood. Shrews (Whitaker and Maser 1976), moles (Whitaker et al. 1979), amphibians, and snakes are the most obvious consumers of detritivores in down wood, while woodpeckers are the most obvious consumers of detritivores in standing dead wood.

A major food item of several carnivores derives from streams and the annual anadromous fish migration. This was no doubt more important in former times, when raptors, such as eagles, condors, and ospreys, and mammals, such as grizzly bear and otter, received major food supplements from this outside source of energy.

The fact that carnivores are so abundant and represent such a critical component of the functioning forest ecosystem will be highly relevant to any conservation strategy as discussed in later sections.

Ordination of Species

During the summer of 1981, we conducted field surveys of fifteen old-growth sites in the H. J. Andrews Experimental Ecological Reserve in the Willamette N. F. (Harris et al. 1982). Restricting our focus to amphibians, reptiles, and mammals, a subjective probability of occurrence was assigned to each species (works by Bratz 1950; Bowles 1963; Anderson 1970, 1972; Stiles 1973; Meslow and Wight 1975; Wiens and Nussbaum 1975; Meslow 1978; Wiens 1978; Mannan 1982; Marcot in prep.; Raphael in prep.; and others treat the avifauna). Three principal site characteristics emerged as most important in determining individual species occurrence as well as the overall species richness of the community. These were (1) presence or absence of surface water and moistness of the site; (2) elevation; and (3) structural complexity of the community. The distributions of all western Cascade amphibian, reptile, and mammal species were then specified along each of these three gradients. Other characteristics, such as size of stand and degree of isolation, were also evaluated but are not reported here.

In establishing the ordinations, it was necessary to distinguish between primary and secondary habitat. Some amphibians, for example, require surface water for the larval life stage. Therefore, even though an adult salamander might occupy a dry site, this could not be considered adequate habitat unless reproduction were possible. Similarly, a mammal such as the marten may use clearcut areas but is not known to meet all of its life functions on these sites (Allen 1982). Primary habitat is therefore distinguished on the grounds that a species can meet all of its life history functions on such a site.

Moistness of site denotes the presence or absence of surface water as well as the mesic to xeric site characterization. By our classification, a "very wet" site is nearly synonymous with aquatic. A wet site would either contain a seep, a pond, or a perennial stream and support vegetation similar to the *Tsuga/Polystichum-Oxalis* or *Tsuga/Polystichum* associations of Franklin and Dyrness (1973, 73–78). On the other extreme, or "very dry" characterization refers to sites with a southwest or south aspect, and a dominance of talus or coarse soils and/or steep slopes. Vegetation would correspond to the *Pseudotsuga/Holodiscus* association of Franklin and Dyrness (1973, 73–75).

Very wet sites serve as primary habitat for 11 mammal and 18 amphibian and reptile species. These sites provide secondary habitat for an additional 8 species (figs. 5.5, 5.6). At least 11 species

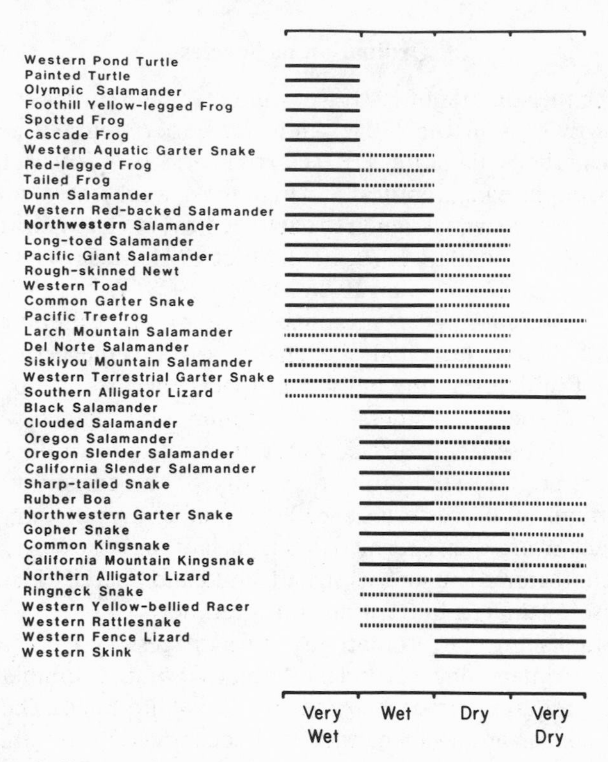

Figure 5.5 Ordination of western Oregon amphibian and reptile species along the moisture gradient. Solid line refers to primary habitat; dotted line refers to secondary habitat.

would not occur if surface water were not present on the site. Wet or mesic sites provide primary habitat for 67 species and secondary habitat for an additional 28 species. Dry sites constitute primary habitat for 57 species and provide secondary habitat for an additional 38 species. Very dry sites provide primary habitat conditions for 30 species while an additional 26 species utilize these sites as secondary habitat. Therefore, whereas 95 species occur frequently on both moist or dry sites, the moist sites serve as primary habitat for nearly 20% more species than the drier sites. Neither the very wet nor very dry sites are ideal for many species.

Elevation was found to be the single most important variable governing the number of species occurring on a site. Of the 108 species of amphibians, reptiles, and mammals inhabiting western Oregon forests, 95 occur at 500 feet (150 m) elevation, while only

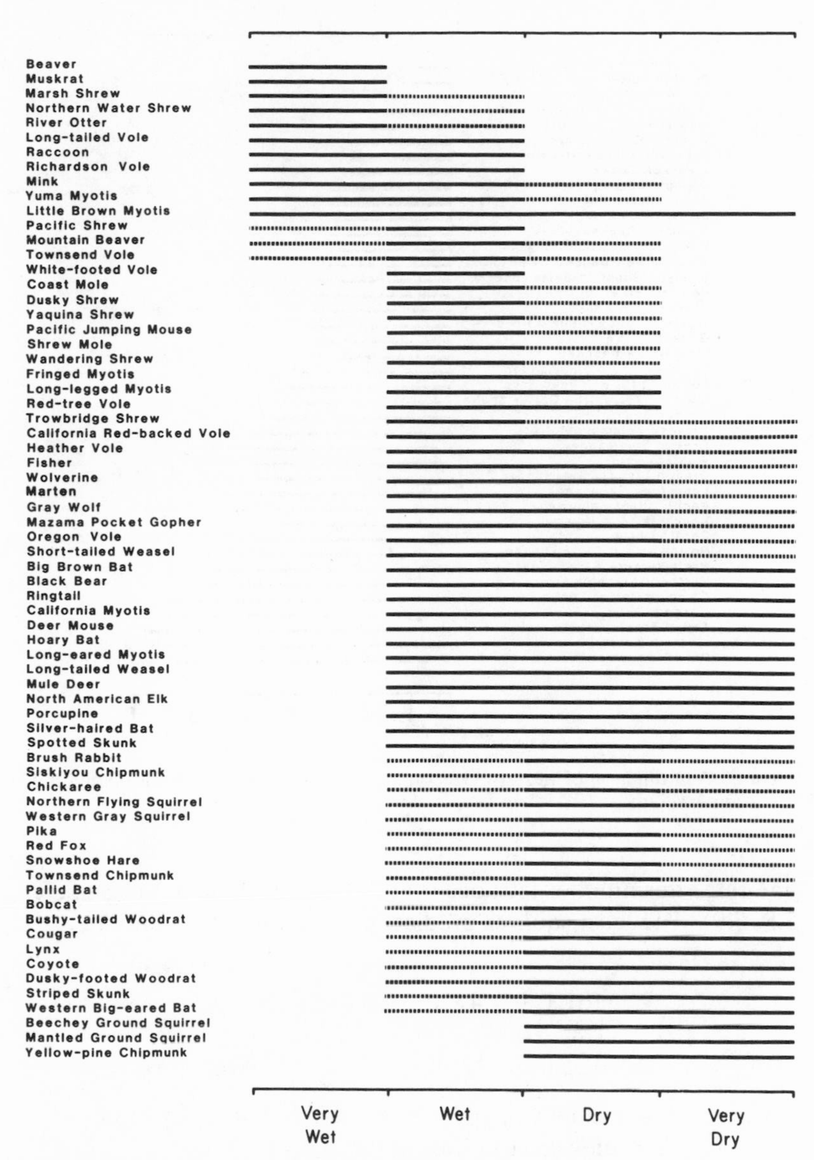

Figure 5.6 Ordination of western Oregon mammal species along the moisture gradient. Solid line refers to primary habitat; dotted line refers to secondary habitat.

Figure 5.7 Ordination of western Oregon amphibian and reptile species along the elevation gradient.

one-third this number (32) occur at 7,000 feet (2,100 m) (figs. 5.7, 5.8, 5.9). The least-squares prediction equation for the number of species is:

$$\hat{S} = 101.1 - 4.4 * 10^{-3}X - 8 * 10^{-7}X^2,$$
$$r = -0.995$$
$$P < 0.01$$

where S = number of species occurring at the elevation
X = elevation in thousands of feet
r = coefficient of correlation

As might be expected, the number of amphibians and reptiles declines rapidly with increasing elevation while the number of mammal species decreases somewhat more slowly. Therefore, when the relative abundance of amphibians, reptiles, birds, and

Figure 5.8 Ordination of western Oregon mammal species along the elevation gradient (figure from Harris et al. 1982).

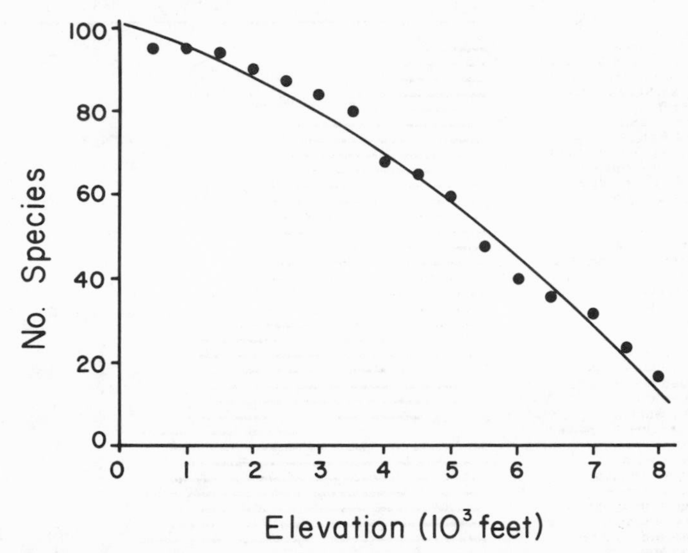

Figure 5.9 Relation between elevation and total number of western Oregon amphibian, reptile, and mammal species. The least-squares equation describing the line is: $S = 101.1 - 4.4 * 10^{-3}X - 8 * 10^{-7} X^2$, $r^2 = 0.99$.

mammals at high elevations is compared to that at lower elevations several trends appear. Amphibians, reptiles, and birds all become relatively less abundant while mammals become relatively more abundant at the higher altitudes (see fig. 6.12). The decline in vertebrates with elevation is correlated with the reduced occurrence of hardwoods, but no quantitative causal relation has been demonstrated. The amount of land area at higher elevations (especially on conically shaped mountains) is also less than the amount of land area at lower elevations and thus a subtle form of species-area relation may be applicable.

Apparently, two distinct shifts occur in the avifauna with increasing elevation. The first shift occurs at the transition between the lower elevation hardwoods and the Douglas fir forests. Species richness, diversity, and abundance are higher in hardwood forests in all seasons than in the Douglas fir forests (Blake 1926; Dirks-Edmunds 1947; Hamilton, 1962; Anderson 1972; Wiens and Nussbaum 1975). Bowles (1963), Anderson (1970), and Wiens and Nussbaum (1975) have observed that whereas the birds of old-growth Douglas fir forests are predominantly permanent residents, many species that occur in hardwood forests or early succession

Douglas fir forests are neotropical migrants. Because the birds of old-growth Douglas fir are predominantly permanent residents, their numbers are more constant throughout the seasons (Bowles 1963).

A different pattern occurs at higher elevations near the alpine zone where winter snow requires that animals either migrate, hibernate, or be small enough to excavate beneath the snow. Mammals are virtually the only permanent residents under such conditions as the birds emigrate to lower elevations (Anderson 1970; Wiens and Nussbaum 1975). Even the birds that occur at the higher elevations during summer are larger, strong-winged insectivores and raptors.

Although the available data are not conclusive, there is reason to believe that the density of individuals of all species combined is also inversely related to elevation. The highest site investigated (4,700 ft = 1,430 m) in the H. J. Andrews Forest supported the lowest density of birds (Wiens and Nussbaum 1975) and this also is suggested by the H. J. Andrews small mammal trapping data (unpublished USIBP data, Arthur McKee personal comm.). Trapping success data for small mammals on Mount Rainier also are negatively correlated with elevation (data from Schamberger 1970, $\hat{y} = 0.106 - 0.01$ thousand feet, $r = 0.37$, $P < 0.01$).

Successional stages were identified and labeled for convenience and as a reference point for discussion. We identified six developmental stages (table 5.4).

Amphibians and reptiles tend to favor the clearcuts and regeneration stands and show little or no attachment to habitat structural characteristics other than decaying logs (fig. 5.10). Mammals re-

TABLE 5.4
Size (diameter breast height) and age values associated with successional stages used for ordination of species.

Stage	Type	Approx. tree size (cm)	Approx. age (yrs.)
1	regeneration	—	0–10
2	seedling-sapling	13	10–50
3	pole timber	25	50–100
4	sawtimber	40	100–150
5	large sawtimber	55	150–250
6	old growth	75	250–450

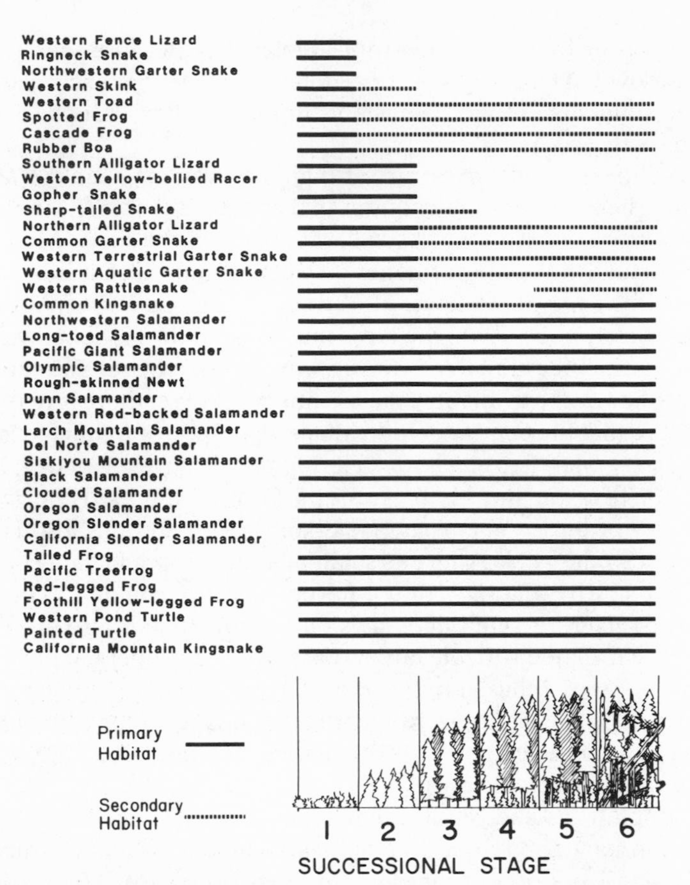

Figure 5.10 Ordination of western Oregon amphibians and reptiles along the successional stage gradient.

spond differently with about equal numbers of species occurring in regeneration stands (58) and old-growth forests (60) (fig. 5.11). It is with regard to mammals, however, that the distinction between primary and secondary habitat becomes critical. Whereas fifty-four mammal species use the pole-timber stage of Douglas fir forests, only 26% of these (fourteen species) find all of their requirements there. Twenty species of mammals find acceptable primary habitat in either regeneration stands or old-growth forests, but not in the mid-successional pole-timber stands. This is probably because the vegetation prevalent in regeneration stands is similar to that occurring as understory in the open patches of old-growth forests. Very

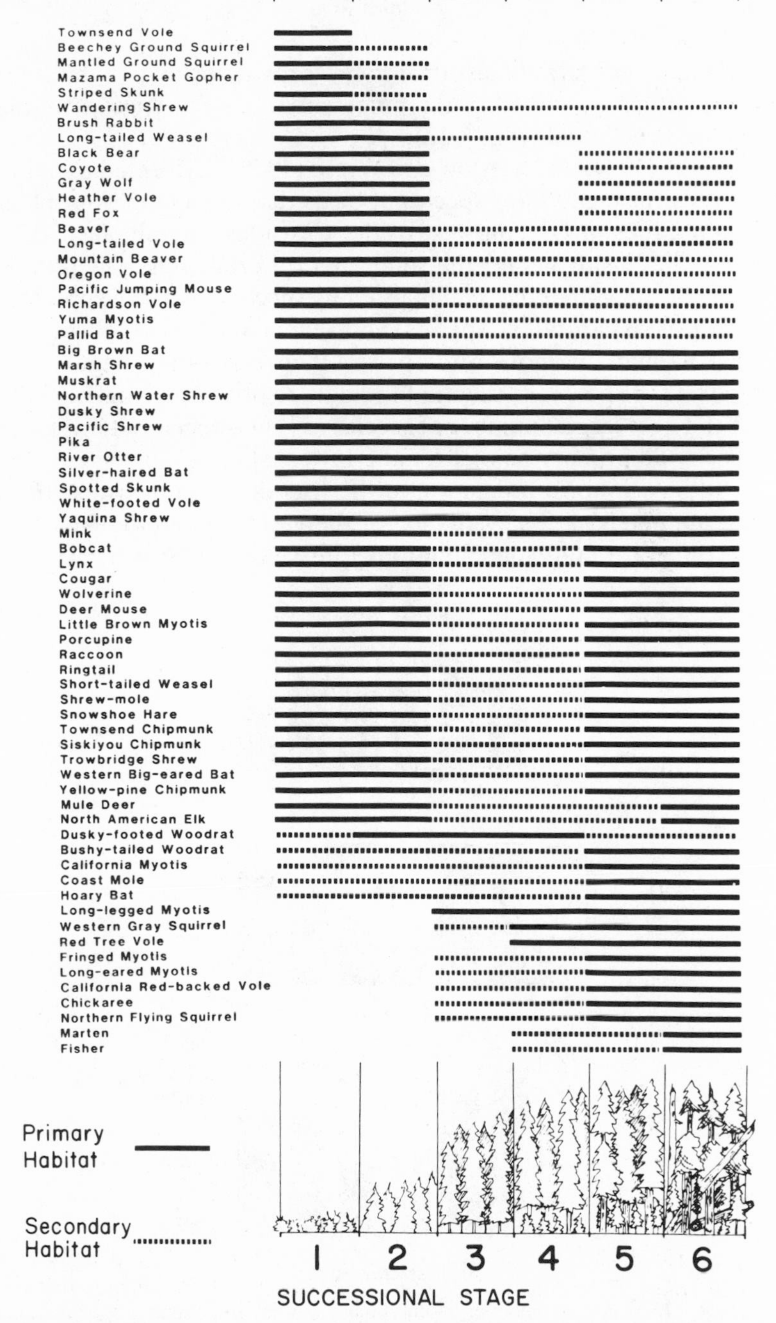

Figure 5.11 Ordination of western Oregon mammals along the successional stage gradient (figure from Harris et al. 1982).

early and very late successional stages provide primary habitat for twice as many amphibian, reptile, and mammal species as the middle-aged stands (fig. 5.12).

Wight (1974) and Meslow and Wight (1975) performed a similar analysis for eighty-four species of birds that use Douglas-fir plant communities in Oregon west of the Cascades summit (fig. 5.13). Making adjustments for the somewhat different stages identified, their list shows that the seedling-sapling successional stage provides both primary (fifty-five species) and secondary (twenty-three species) habitat for more bird species than any other stage. They reported the pole- and saw-timber stages to provide primary habitat for the next largest number of species (54) and regeneration stands to provide primary habitat for the fewest species (6).

Adding the bird ordination to the lists of amphibians, reptiles, and mammals allows evaluation of the various successional stages for all western Oregon coniferous forest vertebrate species (table

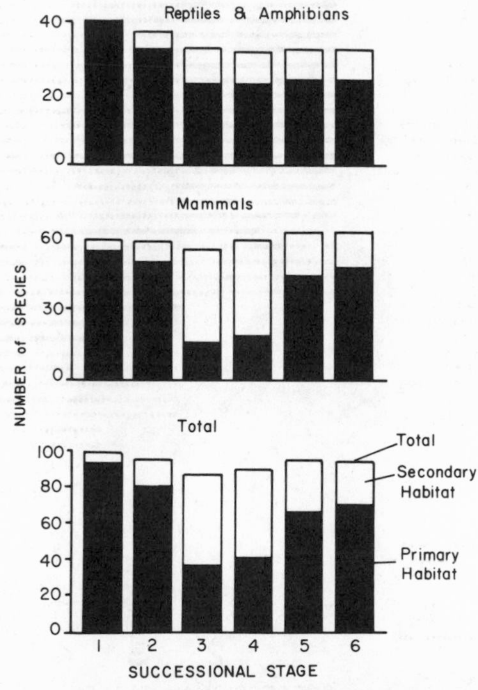

Figure 5.12 Combined number of amphibian, reptile, and mammal species occurring in each of six stages of the Douglas fir successional sequence (figure from Harris et al. 1982).

5.5). Considering all species that commonly use the various stages for either primary or secondary habitat, there is a high degree of constancy around 153 species. Regeneration stands (stage 1) are inhabited by seventeen fewer species (136), whereas seedling and sapling stands (stage 2) are inhabited by nineteen more species (172). But the crucial question from a wildlife habitat standpoint is,

Figure 5.13 Ordination of the eighty-four species of birds associated with various successional stages of western Oregon coniferous forests. A single line represents "occurs there"; a double line represents "nests there"; a triple line represents "primarily nests there" (modified from Meslow and Wight 1975).

TABLE 5.5
Number of terrestrial vertebrate species using the six successional stages for primary habitat (acceptable for escape, feeding, reproducing, and overwintering) and for secondary habitat.

	Successional stage					
	1	2	3	4	5	6
Primary habitat	99	135	90	93	114	118
Secondary habitat	37	37	61	60	39	35
Total no. species	136	172	151	153	153	153

for how many species do the successional stages provide primary habitat? These statistics reveal a different pattern. Again, seedling and sapling stands provide primary habitat for the most species (135) with old-growth and mature forests serving as primary habitat for about 15% fewer species and pole-timber stands supporting 33% fewer species than the seedling-sapling stands.

Richness vs. Diversity

The analysis above has dealt with species richness of different-aged stands located on different types of sites. It must be kept in mind that the ordinations of species gradients were based on natural history information of the individual animal species and not on measurements from actual plots. This means that in all cases we implicitly referred to potential species richness and not measured species richness.

A second point of great importance is the distinction between species richness and species diversity. The species diversity of a site should be thought of as the probability that two randomly selected organisms belong to the same habitat but different species (Patil and Taillie 1982). This means that relative abundance of the different species must be taken into account. Obviously if a few species are very abundant and totally dominate the community, then the probability that a randomly selected sample will consist of all different species will be low. Both the perceptual and measured diversity will be low. In this sense, diversity is indeed a measure of average rarity within the community (Patil and Taillie 1982).

Referring back to the successional data cited above, it will be noted that stages 1 and 2 of the successional sequence provide primary habitat only for species that are common in second-growth throughout the western Cascades. In other words, all of the species that meet their primary habitat requirements in these early stage forests find abundant habitat throughout the western Cascades and are generally common. Forty of the species finding primary habitat conditions in old-growth or mature forest cannot meet their habitat requirements outside this forest type.

Bowles (1963) used frequency of observation to rank different species of birds into common ($> 21\%$), fairly common (11%–20%), and rare ($< 11\%$) categories. A highly significant interaction exists between the commonness of species and the plots in which they occur (table 5.6). Whereas twice as many "common" species occurred in a salvage-logged blowdown area as occurred in the old growth, twice as many "rare" species occurred in the old growth as occurred in the salvage-logged area (table 5.6). The most frequently seen species in cutover areas were summer residents, but the most frequently seen species in the old growth were permanent resident species (Bowles 1963).

A recent study of avian community diversity in Douglas fir forests of northeastern Oregon revealed yet other results. Mannan (1982) found that the density of breeding birds was greater in recently thinned, rotation-age forest stands ($\simeq$ 85 years old) than it was in 200-year-old stands. However, the number of breeding bird species, the evenness with which individuals occurred within species, and the overall species diversity was greater in the 200-year-old stands. This result derived primarily from the absence of three species (varied thrush, Swainson's thrush, and goshawk) and the much lower density of eight species (Vaux's swift, pileated woodpecker, Williamson's sapsucker, gray jay, brown creeper, Townsend's warbler, golden-crowned kinglet, and hermit thrush) in the

TABLE 5.6

Number of species of birds ranked as common, fairly common, and rare occurring in the three types of Douglas fir forests in western Washington (from Bowles 1963).

	Common	Fairly common	Rare
Salvage logged blowdown	16	12	8
Partially logged mature stand	15	5	11
Old growth	7	5	17

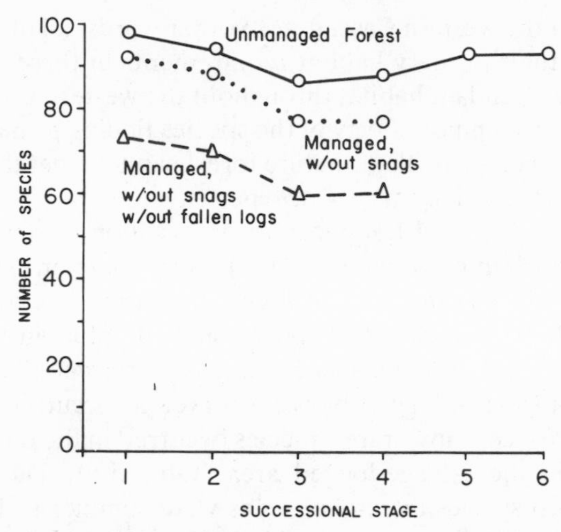

Figure 5.14 Number of terrestrial vertebrate species (excluding bats) potentially occurring in an unmanaged Douglas fir forest and in two generations of managed Douglas fir forests in western Oregon (figure from Harris et al. 1982).

younger-aged stands. Six species (Hammond's flycatcher, dusky flycatcher, ruby-crowned Kinglet, Cassin's finch, dark-eyed junco, and chipping sparrow) were more abundant in the younger stands. In general, hole-nesting and bark-foraging birds were more abundant in the 200-year-old stands.

The two habitat elements that contribute most to the habitat value of older forests are standing large, dead trees (snags) and fallen logs (Maser et al. 1979; Thomas et al. 1979a; Franklin et al. 1981). First-generation plantations where residual snags and logs have been retained will provide habitat for species that will not occur in later-generation plantations not containing these elements. The tally of species (excluding birds) that will occur in short-rotation stands without snags is 10% below the number occurring in stands with snags (fig. 5.14). The number of species occurring in short-rotation stands that contain neither snags nor fallen logs is about 29% below the number occurring in unmanaged old-growth stands.

Short-rotation forests that do not contain a significant percentage of mature and old-growth forest with abundant snags, fallen logs, broken-top trees, and cavities will not provide primary habitat for forty-five species of terrestrial vertebrates (appendix 4). Although several of these species may use short-rotation forests as secondary habitat, they require that older-age stands or specific patches of primary habitat be present somewhere in the forest.

TABLE 5.7

Distinctive wildlife community characteristics of old-growth Douglas fir ecosystems of western Oregon (and western Washington).

- Because of the great diversity of ecosystem types, western Oregon supports more bird families than any other area in North America. It also supports many more breeding bird species and resident mammal species than most other areas. Fewer amphibian and reptile species occur relative to eastern and southeastern regions.

- Mature and old-growth coniferous forests support the only two folivorous species occurring in North America. The red tree vole is the most highly specialized vole in the world and the most arboreal mammal in North America. The white-footed vole is the smallest mammalian browser and perhaps the rarest vole in North America. Both species are endemic to these forests and occur nowhere else in the world.

- The only fungivorous mammal in North America occurs in mature and old-growth coniferous forest. This seems to be a case of coevolution wherein the hypogeous (below ground) fungi have adapted to animals for spore dispersal while the California red-backed vole has coadapted to the fungi as its major food source. This vole is endemic to these forests and occurs nowhere else in the world.

- In the absence of any large-seed-bearing tree species, the northern flying squirrel exhibits a strong dependence on epiphytic foliose lichens and fungi as its food source. It is the only species in North American forests whose winter diet is predominantly lichens.

- The carnivore species richness is great but few herbivore species, in the strict sense of the word, exist in the coniferous forest. This results in 65% of the terrestrial vertebrate species being carnivores and a dependence on the folivore, fungivore, detritivore, lichen-eating, and riparian stream energy pathways for support.

- The vertebrate fauna consists of approximately 58% birds, 30% mammals, 7% amphibians, and 5% reptiles. All of the amphibians, reptiles, and mammals are permanent resident species and as high as 75% of the birds are permanent residents. This is unlike the eastern forest, where about 75% of the aviafauna is migratory.

- Within the coniferous forest type, three environmental gradients govern vertebrate species richness. These are elevation, presence or absence of water and moistness of site, and structural diversity, which is related to successional stage. Other factors, such as the presence or absence of landscape "benches" on the site, area encompassed by the stand, and presence of hardwoods are also important determinants of species richness and habitat quality.

- Along the moisture gradient, primary habitat is provided for most species on moist sites containing surface water, preferably a riparian strip of hardwood and a topographic bench.

- Lower-elevation sites provide primary habitat for several-times more species than higher-elevation sites. Existing data suggest that the density of all species combined is also greater at lower elevations.

TABLE 5.7 (continued)

- Old-growth and mature forests provide primary habitat for 118 species of vertebrates, over one-third of which (40 species) cannot find primary habitat outside this forest type. Although the seedling-sapling stage of forest succession provides primary habitat for more species (135), many of these species are very common throughout the region in suburban, agricultural, and ruderal habitats, and forest management is unnecessary for their well-being.

- Standing dead trees, broken top and decaying live trees, and fallen logs are principal structural components of old-growth forest habitat. Forty-five species of terrestrial vertebrates will not occur in young forests that do not contain these structural components.

PART 3

Analysis of Alternatives

6

The Applicability of Insular Biogeography

Prior to developing scheduling strategies based on insular biogeography principles, it may be useful to review basic concepts as they apply to the Cascades and the Northwest in general. I will not cite the scores of published articles debating the specific issues since these can be quickly reviewed elsewhere (e.g., F. S. Gilbert 1980; Simberloff 1982; Simberloff and Abele 1982). I will generally limit citation of data papers to those deriving from western and northwestern studies.

Two Types of Islands

Both Darwin and Wallace distinguished between continental shelf islands recently isolated by rising sea level and oceanic islands whose origin was independent of any continental connection (Sauer 1969). Aside from individual characteristics that the two types might exhibit, the origin and trend of species numbers across these types warrant discussion.

Oceanic islands (or continental volcanic cinder cones) begin with barren substrate and the animal community develops from a few initial colonists to progressively richer and more complex levels. Immigration and colonization rates, along with the factors that control these rates, predominantly influence the characteristics of the biota of such islands, at least initially.

In contrast to oceanic islands, islands such as the Aleutians were formerly points on a continuous landmass. The animal communities of these islands have regressed from higher levels of species richness characteristic of larger continents to a reduced number of species characteristic of islands (fig. 6.1). Even when standardized for size and degree of isolation, true oceanic islands generally do not develop the same level of species richness as continental shelf islands (Darlington 1957; MacArthur and Wilson 1967). To the extent that

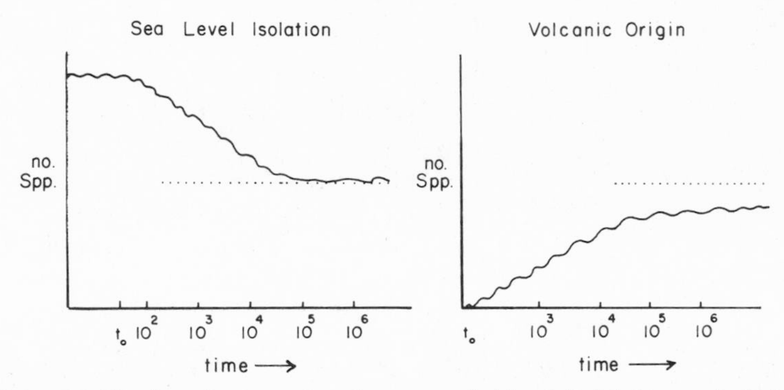

Figure 6.1 The faunal community of true oceanic islands builds from a single first colonist to some equilibrium level. The fauna of continental islands or residual habitat islands consists of a subset of some richer original community after relaxation has occurred. Even when standardized for size and degree of isolation, oceanic islands probably never reach the same level of richness as continental islands.

the true island analogy applies to forest habitat islands, this phenomenon has direct implications for old-growth management that will be discussed in a later section.

Since both of the processes described above require time for their effects to be measured, time since origin is another critical variable influencing characteristics of the plant and animal communities. Two examples will demonstrate the principle and process. Hickman (1968) surveyed the vegetation of forty-two mountain peaks in the Oregon Cascades and analyzed the data with respect to several environmental variables. Focusing just on disjunct species (those that do not have continuous distributions with other regions), he found that recent volcanic cones (e.g., Sand Mountain, Hoodoo Butte) supported an average of forty-five species. Older peaks (e.g., Maxwell Butte, Grizzly Peak) with more developed substrates and a much longer time for invasion supported an average of 130 species. These statistics illustrate the principle that the amount of time since creation is a determinant of species richness when dealing with montane peaks of volcanic origin that are analogous to true oceanic islands.

Evidence for the second pattern of faunal change derives from areas such as Mount Rainier National Park. Early surveys report that in 1920 fifty species of mammals were present in the park (Taylor and Shaw 1927). By 1935 the number of species appears to have been reduced to forty-nine (Kitchen 1935, in Weisbrod 1976), and by 1976 the number of species was only thirty-seven. The loss of thirteen species (26%) has apparently occurred in spite of the invasion of the park area by several new immigrant species (e.g., porcupines, striped skunk) (fig. 6.2). After analyzing mammal distributions on Great Basin montane islands, Brown (1978, 223) concluded that "in the absence of immigration they [island faunas] are gradually relaxing toward an equilibrium of zero species at rates inversely related to island size." Grayson (1982) provides considerable support for this and related conclusions. It has been pointed out elsewhere that these trends are not inevitable, but simply represent the natural process most likely to occur in the absence of mitigative action (Miller and Harris 1977).

In the case of old-growth habitat islands, time since origin can affect faunal conservation in several ways. Based on the analogy that continental or landbridge islands isolated from a larger species pool ultimately support more species than comparable oceanic islands, old-growth islands that can be salvaged from existing large tracts will be superior to those derived from the development of isolated replacement stands. Development of strategically located young stands will be necessary in certain instances, but this

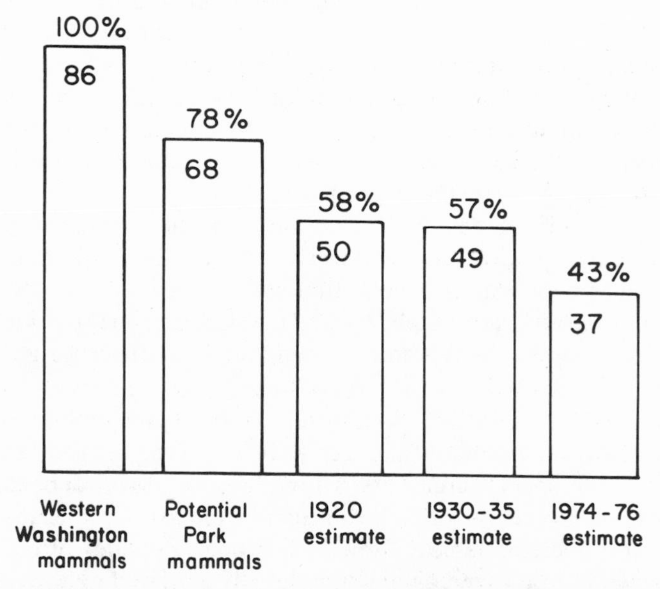

Figure 6.2 Number of mammal species occurring in western Washington and in Mount Rainier National Park at various times in recent history (from Weisbrod 1976).

approach is not equivalent to the protection of remnant old-growth stands.

When conservation of isolated remnant old-growth islands has been achieved, "relaxation" toward a smaller species number is to be expected even if the old growth maintains its same general character in perpetuity. But since the old-growth ecosystem will change in its own right, wildlife species may be further reduced. This implies that the ideal strategy involves conservation of remnant old-growth stands, provision of adjacent replacement stands, and provision for reinvasion by species that have become locally extirpated from the stands.

Species-area Relations

Grinnell and Swarth (1913) were perhaps the first to draw attention to the relation between the number of species and the area of montane peaks. They went so far as to state: "A possible law appears justified: the smaller the disconnected area of a given zone . . . the fewer the types which are persistent therein" (Grinnell and Swarth 1913, 385). Attempts to establish quantitative relations between area and number of species were debated by Arrhenius (1921, 1922) and Gleason (1922, 1925). Both the concept and its utility were appreciably advanced by Cain (1938), Vestal and Heermans (1945), Vestal (1949), and Hopkins (1955). Although efforts to develop quantitative relations were all directed at terrestrial communities, the same issue was being addressed by students of oceanic biogeography (Darlington 1957).

Given any homogeneous biotic community, the average number of species per quadrat sample will increase as the quadrat size is increased. When graphed on arithmetic paper, a typical species area curve results (top graph, fig. 6.3). When graphed on double logarithmic paper, the relation is essentially linear (middle graph, fig. 6.3).

Preston (1962a, 1962b), MacArthur and Wilson (1967), and others have stressed the differences to be expected between parameters of the species-area curves for the same taxon depending upon whether the data derived from sample plots or true "islands." For quite understandable reasons, the slope (Z-value) of curves deriving from true island data is significantly greater than the slope of curves deriving from sample plots and continental habitat islands (lower tier, fig. 6.4). The magnitude of and changes in these slope

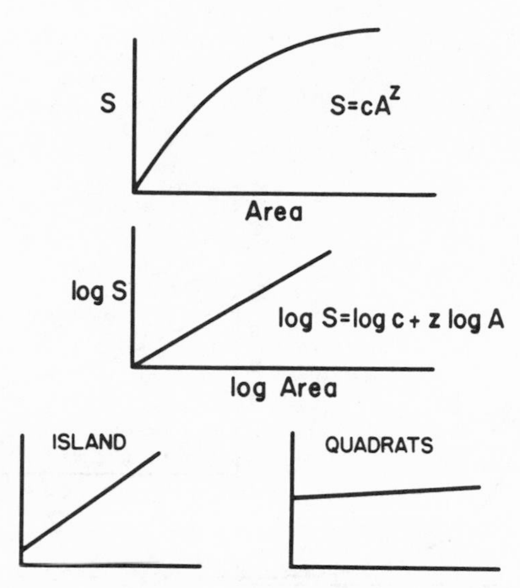

Figure 6.3 Increase in number of species occurring on progressively larger-sized sample plots or islands. The allometric relation shown in the top frame is approximately linearized by a log-log plot of the data. Isolated islands generally exhibit steeper slopes than sample plots surrounded by similar habitat.

values have been cause for considerable debate (Connor and McCoy 1979; Martin 1981; Sugihara 1981).

One of the most fruitful lines of inquiry deriving from species-area analysis is the biological basis for cause-and-effect relations and the implications for resource management. Consider, for example, the biogeographic axiom, "an island of 10 times greater size is required to support a fauna of two times as many species" (Darlington 1957). The converse of this is that the area of an island may be reduced by 90% while the faunal species richness is halved. This relation does in fact hold if the slope coefficient of the double logarithmic plot is 0.30. But the curve is only a manifestation of the power function

$$S = cA^z$$

When variables appear as exponents, their values have dramatically powerful effects on the dependent variable, in this case number of species. For comparison, let Z take the value 0.5, in which case only a fourfold increase in area is necessary to double the number of species. On the other hand, when Z takes a lower value such as 0.14, then the size of the area must be increased 140-fold to double

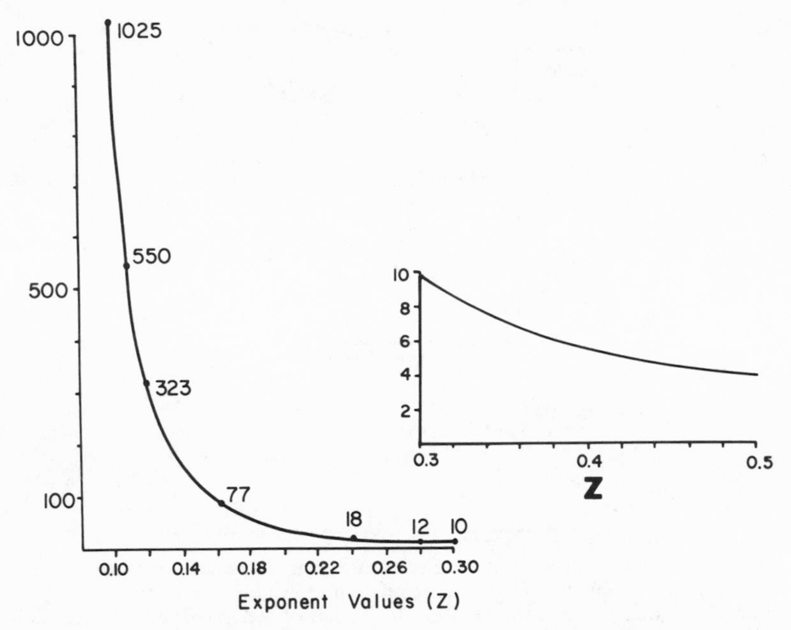

Figure 6.4 Relation between the value of the exponent Z in the species-area equation $S = cA^z$ (see figure 6.3) and the increase in area that would be necessary to double the number of species present. For example, if the Z-value were 0.30, then a tenfold increase in area would be necessary to double the number of species.

the number of species (fig. 6.4). Therefore, in the absence of empirically derived species-area curves for old-growth islands, generalities must be used with caution. On the other hand, the hypothesis that distinctly isolated old-growth habitat islands will manifest a species-area relation with a slope of approximately 0.35 is clearly falsifiable and should be tested.

Since the early work of Grinnell and Swarth (1913) and Grinnell (1914a, 1916) in California, descriptions of depauperate faunas occurring on montane forest islands surrounded by a "sea" of drier grassland or desert have been published for the Sierra del Carmen of Mexico (Miller 1955), the Mogollon Mountains (Hubbard 1965), the Sandia Mountains of New Mexico (Tatschl 1967), the Spring and Sheep Ranges of Nevada (Johnson 1965), several ranges in the northern Great Plains (Thompson 1974), and dozens of sample areas in the Sierra Nevada and Rocky Mountains (Brown 1971, 1978; Johnson 1975; Honacki 1978; Picton 1979) (fig. 6.5). Studies by Gavareski (1976) and Willson and Carothers (1979) establish species-area relations for breeding birds in forested urban parks in

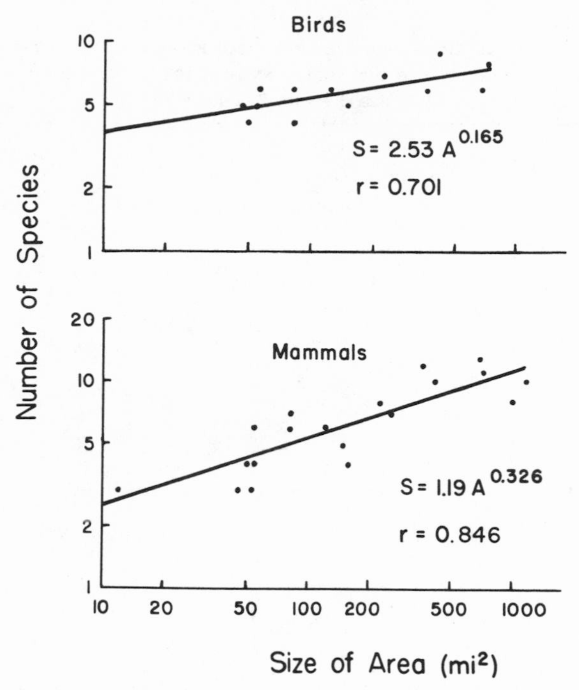

Figure 6.5 Relation between number of permanent resident boreal bird species, the number of small, boreal mammal species, and the size of isolated mountain ranges (>7,500 ft, 2,300 m elev.) in the Great Basin (modified from Brown 1978).

Seattle and for habitat islands in the Grand Canyon respectively. All of these studies lend support to the principle that animal species richness increases with size of area.

Bratz (1952) drew attention to the "anomalous" biota of several areas in Oregon and concluded that the size of isolated areas such as Mary's Peak was a major factor governing the biotic characteristics. Bekele (1980) reported a significant positive correlation between the number of resident large mammal species and the size of area for fourteen national parks in the western United States (table 6.1). As might be expected, Thompson (1974) found that area effects were more pronounced for permanent resident birds but much reduced for migrant species. Because the coniferous forests of the western Cascades support predominantly permanent resident bird species, one might expect strong area relationships to prevail. Although only a few of the studies cited were sufficiently quantitative to establish a specific species-area relation, there remains little

TABLE 6.1

Number of species of large mammals in western national parks. Asterisks denote parks of special relevance to this study; areas of the parks are given in square kilometers (from Bekele 1980).

	Bryce Canyon (144)	Redwood (250)	*Lassen Peak (426)	Zion (588)	*Mount Rainier (968)	Rocky Mountain (1049)	*North Cascade (2022)	Big Bend (2833)	Yosemite (3044)	Sequoia/King's C. (3366)	Grand Canyon (3487)	*Olympic (3586)	Glacier (4053)	Yellowstone/Grand T. (1,012,900)
javelina								X						
pronghorn								X			X			X
mountain goat					X	X	X					X	X	
bighorn sheep				X		X			X	X	X		X	X
black-tailed deer		X												
mule deer	X		X	X	X	X	X	X	X	X	X	X	X	X
white-tailed deer								X					X	X
Roosevelt elk		X										X		
elk	X			X	X	X	X				X		X	X
moose							X							X
coyote	X	X		X	X	X	X	X	X	X	X	X	X	X
gray wolf							X						X	X
bison														X
mountain lion	X	X		X	X	X	X	X	X	X	X	X	X	X
black bear		X	X	X	X	X	X	X	X	X	X	X	X	X
grizzly bear													X	X
No. of species	4	5	2*	6	6*	7	8*	7	5	5	7	6*	10	12

doubt that the general relationship holds for the western Cascades. It is difficult to establish any single set of parameter values when different taxa are evaluated under different environmental conditions by varying techniques and standards. On the other hand, when the taxa and standards for defining a "resident species" have been specified, then region-wide species-area relations probably do exist.

The use of a parameter such as area for predicting species richness may be criticized on grounds that it is too simplistic and obscures the true mechanisms that actually determine diversity (Connor and McCoy 1979). Brown (1978) and others have addressed this point and there is no need to repeat the discussion here. Suffice it to say that heterogeneity within the old-growth stand is important, as are all of the characteristics discussed in the previous chapter. Clearly some old-growth ecosystems will support more species than others. However, this does not detract from the fact that a statistical relation between size and richness exists.

As the species richness of a community increases, patterns in the characteristics of colonizing species appear (Diamond 1975). Thus, as the number of species in an area increases, the probability that two species with similar food and habitat requirements will co-occur also increases. This means that interspecific competition is likely to be higher. Similarly, as more species are added to the community, the number of predator species increases and the role of predation as a limiting factor increases relative to food limitation (fig. 6.6; Heaney 1978).

When standardized for weight, carnivores and insectivores have larger home-range sizes and area requirements than herbivores, granivores, or omnivores. Thus insectivores and carnivores may be able to reside only in areas sufficiently large to meet their food and area requirements. In part because of this, the proportion of total species that is insectivorous or carnivorous tends to increase as size of area increases. Martin (1978) reported such a relationship for

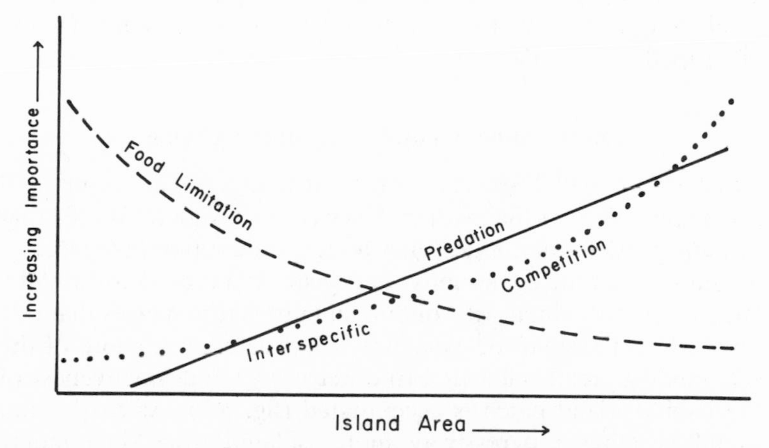

Figure 6.6 Trends in three ecological parameters believed to occur in response to island size and/or the process of isolating a habitat island fauna (modified from Heaney 1978).

TABLE 6.2
Percentage of bird species occurring in small-, medium-, and large-midwestern habitat islands that are granivores, omnivores, and insectivores. Numbers represent two-year averages (from Martin 1978).

	Habitat island size		
	small	medium	large
% Granivore	34	28	22
% Omnivore	32	32	30
% Insectivore	36	42	48

Dakota forest habitat islands (table 6.2), and the relationship probably holds for the western Cascades also. Smaller islands can be expected to support herbivores, granivores, and omnivores in reasonable numbers but larger islands are required for many insectivores and carnivores.

An additional generality applies, however. The habitat specificity (as we generally define it) of animals tends to be inversely related to area requirements. Thus, large, wide-ranging resident species such as cougar tend not to have specific habitat requirements. Conversely, species such as California red-backed vole or the red tree vole that have highly specific habitat requirements do not range widely. This means that the largest, widest-ranging species will not be obligatorily linked to "old-growth" islands per se. They certainly would not be expected to reside totally within, or be contained by, old-growth habitat patches.

Local Extinction and Community Change

The term "island" spans the gradient from a small, distant atoll surrounded by a hostile medium (the sea) to a wilderness area that is defined by nothing more than a legal description in legislative minutes. A natural community that occurs as a central portion of a larger regional habitat will contain numerous rare species that rely on the larger system for existence. As progressively more of the surrounding area is allocated to other uses, the distinctiveness of the habitat island patch is accentuated (fig. 6.7). As the habitat island becomes progressively more isolated from surrounding vegetation of similar form, the rare species are quickly lost. The changes described for Mount Rainier National Park and parks and

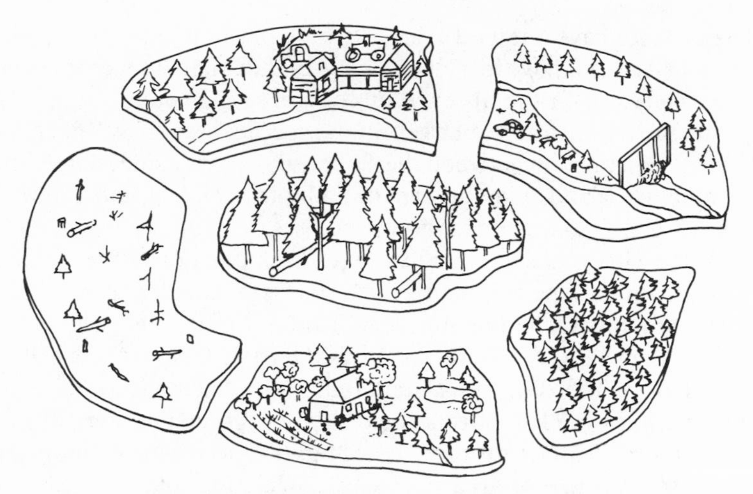

Figure 6.7 Insularization of old-growth ecosystem patches results from numerous forms of forest management and development.

reserves in other areas (e.g., Miller and Harris 1977) result in a depleted fauna with identifiable characteristics (fig. 6.8).

Detailed analysis has led Brown (1978) to conclude that even among small mammals, the "faunas of the montane islands have been derived by extinction from a common set of 14 functional species. . . . Extinction reduced the faunas [sic] of each island and thus played a major role in determining species composition. Five

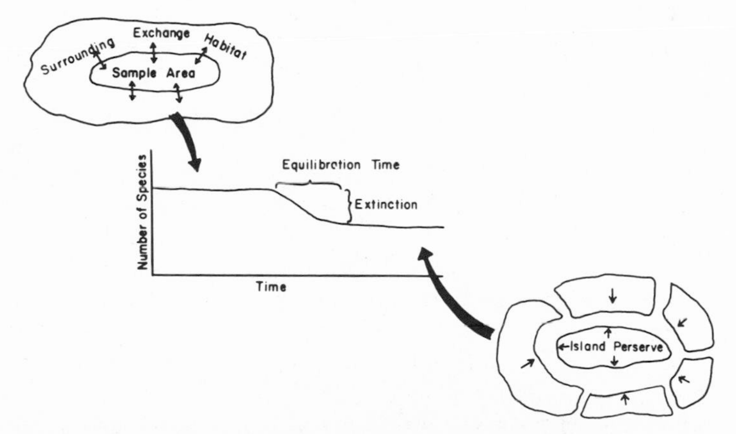

Figure 6.8 As old-growth habitat patches become isolated from similar surrounding habitat, species dependent upon the larger area are lost from the island and the total species number is reduced. (From Miller 1978).

large islands have retained at least 10 of their original 14 species, but five small islands have lost all but 3 or 4 of the original set of species" (Brown 1978). In the absence of compensating colonization, this process results in a steeper slope in the species-area curve (fig. 6.9). Thus the distinction between the slope values of oceanic islands and continental islands may be explained by the relative imbalance of the extinction and colonization processes.

Picton (1979) analyzed the changes in populations of ten native large mammal species on twenty-four semi-isolated continental mountain ranges in the northern Rocky Mountains. The areas varied from 11 to 4,480 square miles (29 to 11,600 km^2). It will be found from the data presented that the percentage of original species that were lost during the period of settlement, agricultural and ranch development, subsistence hunting, logging, and mining is inverse to the size of area (fig. 6.10). While the smaller areas lost over 50% of their original species, the larger areas lost as few as 4% of the species historically present (Picton 1979). In other words, factors that increase the disparity between extinction and colonization cause the slope of the species-area relation to increase toward the values exhibited by true islands.

Local extinction or loss of species from parks, preserves, and habitat islands appears to affect certain groups of species more than others. In general, the species most vulnerable to extirpation are those represented by small populations. Small populations may

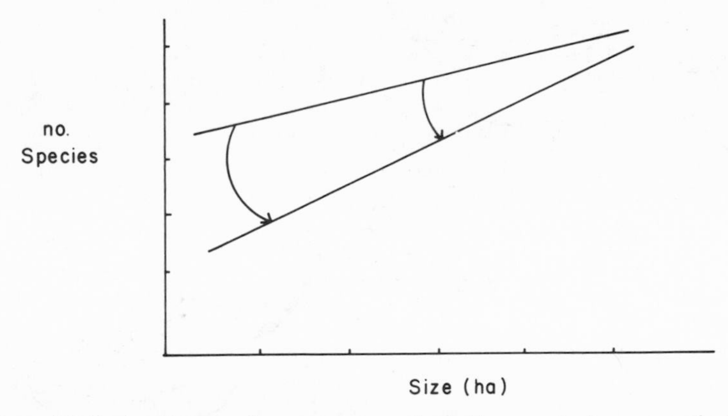

Figure 6.9 In the absence of compensating colonization, species become extinct in small islands faster than in larger islands. Lower colonization rates also explain why distant islands support fewer species than closer islands, and why true oceanic islands have steeper-sloped species-area relations than continental islands.

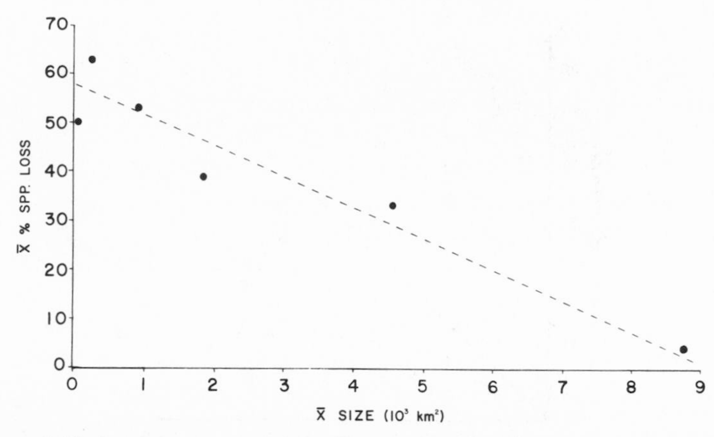

Figure 6.10 Average percentage loss of large mammal species as a function of island size for isolated ranges in the northern Rocky Mountains (data from Picton 1979).

result from temporal and spatial resource limitation (e.g., highly specialized habitat or food source), overexploitation, or consideration of an area that is small relative to the home-range size of the species. Species that are highly specialized, those that represent the largest members of their feeding class (e.g., raptor, terrestrial carnivore, bark-gleaning insectivore), and those that occur at the higher trophic levels are usually the first to be lost. Even though wide-ranging species may have higher recolonization potential, their local extinction rates are higher than average for all species in the community (fig. 6.11).

For various reasons, the phenomenon described above is especially critical for mammals. The grizzly, gray wolf, and fisher have all been extirpated from western Oregon and wolverines and lynx are very rare. Notably, these are all top carnivores. Olterman and Verts (1972) reviewed the status of forty-one Oregon mammal species of doubtful status. Seven of the eight species (88%) ever occurring in the western Cascades and judged to be "extirpated," "rare," or "endangered" are carnivores. Conversely, five of eight species (62%) in the "not rare or endangered" category are herbivores. Picton (1979) restricted his analysis of Rocky Mountain montane areas to large herbivores because the carnivores had been so seriously depleted as to render analysis unwarrantable. In the Great Basin, "herbivorous species of generalized habitat requirements and small to intermediate body size have persisted on most of the islands, . . . herbivores of large body size and/or specialized

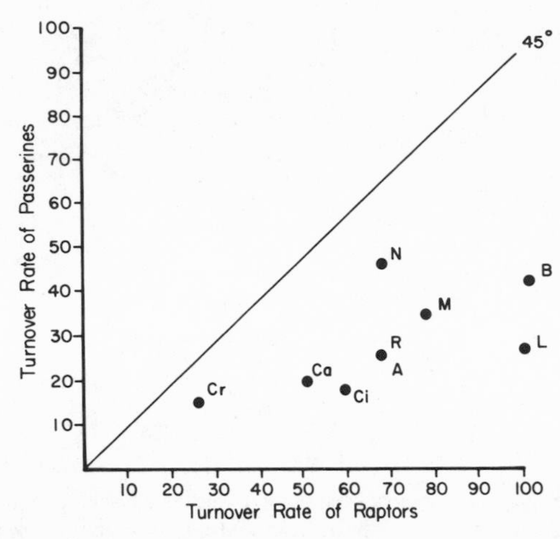

Figure 6.11 Comparative turnover rates for birds of prey versus song birds for eight of the Channel Islands off the coast of California. If raptors and song birds had equal extinction rates, then the data points would fall on the equilibrium line no matter what the specific rate was (data from Hunt and Hunt 1974).

habitat requirements and carnivores have had higher extinction rates and persist on only a small proportion of the 19 islands" (Brown 1978).

Isolation Effects

The importance of isolation in determining the characteristics of biotic communities is one of the longest-standing principles of biogeography (Darlington 1957; MacArthur and Wilson 1967; Pielou 1979). Although distance immediately comes to mind as the principal measure of isolation, the concept is considerably more complex. An absolute distance measure may be convenient for resource managers, but it bears little resemblance to the reality that different plant or animal species face. There is approximately a thousandfold difference in the normal month-to-month travel distances between small mammals such as moles and large ones such as grizzlies and wolves. Therefore, while a 0.6 mile (1 km) distance may be seventy-five times the normal "cruising radius" of a mole, it represents but 1% of the "cruising radius" of a cougar (predicted cruising distances are given in table 6.3).

The second reason that absolute distance measures are inadequate involves the sedentary versus migratory habits of different

TABLE 6.3

Calculated home-range sizes and linear travel distances assuming circular and elliptically shaped home ranges. Shape of the ellipse determined by major axis = 2 × minor axis. 1,000 m = 1 km, 1,609 m = 1 mile.

Species	Home range (ha)	Circular home range		Elliptical home range major axis (m)
		r (m)	2r (m)	
white-footed vole	0.049	12	25	35
Oregon vole	0.053	13	26	37
California red-backed vole	0.056	13	26	38
deer mouse	0.056	13	26	38
heather vole	0.062	14	28	40
Pacific jumping mouse	0.064	14	29	40
red tree vole	0.082	16	32	46
yellow-pine chipmunk	0.10	18	36	50
long-tailed vole	0.15	22	44	62
Mazama pocket gopher	0.17	23	47	66
Townsend vole	0.18	24	48	68
Townsend chipmunk	0.19	25	49	70
Siskiyou chipmunk	0.19	25	49	70
Richardson vole	0.22	26	53	75
wandering shrew	0.28	30	60	84
Trowbridge shrew	0.28	30	60	84
northern flying squirrel	0.33	32	65	92
pika	0.36	34	68	96
northern water shrew	0.44	37	75	106
shrew mole	0.44	37	75	106
dusky shrew	0.44	37	75	106
dusky-footed woodrat	0.48	39	78	111
mantled ground squirrel	0.50	40	80	113
chickaree	0.54	42	83	117
Yaquina shrew	0.56	42	84	120
Pacific shrew	0.68	46	93	132
bushy-tailed woodrat	0.80	51	101	143
marsh shrew	1.09	59	118	167
Beechey ground squirrel	1.34	65	131	185
western gray squirrel	1.76	75	150	212

TABLE 6.3 (continued)

Species	Home range (ha)	Circular home range r (m)	Circular home range 2r (m)	Elliptical home range major axis (m)
brush rabbit	1.76	75	150	212
snowshoe hare	2.55	90	180	255
mountain beaver	2.55	90	180	255
muskrat	2.90	96	192	272
coast mole	3.97	112	224	318
short-tailed weasel	5.26	129	258	370
spotted skunk	26.7	292	584	824
ringtail	30.2	310	620	876
long-tailed weasel	30.7	313	626	884
porcupine	34.9	333	667	942
beaver	53.2	412	824	1,160
striped skunk	101	567	1,130	1,600
red fox	62	718	1,440	2,040
marten	215	827	1,650	2,340
mink	52	896	1,790	2,530
mule deer	420	1,160	2,320	3,270
coyote	453	1,200	2,400	3,400
raccoon	480	1,240	2,480	3,500
elk	943	1,730	3,460	4,900
fisher	1,610	2,260	4,520	6,400
black bear	1,760	2,370	4,740	6,700
otter	3,010	3,100	6,200	8,760
wolverine	4,900	3,950	7,900	11,200
lynx	5,710	4,260	8,520	12,100
bobcat	11,600	6,080	12,200	17,200
cougar	49,700	12,600	25,200	35,600
gray wolf pack (5)	153,000	22,100	44,200	62,400
grizzly bear	377,000	34,600	69,300	98,000

species. With like habitat requirements, a migratory species whose members travel thousands of kilometers seasonally has a higher probability of colonizing isolated islands within the flyway than sedentary permanent residents. For species such as the pika, a distance of 1,000 feet (300 m) between talus outcrops is a difficult barrier and dispersal over a distance of a few kilometers of hostile environment is virtually impossible (Smith 1974).

A third consideration involves habitat specificity and tolerance for variation along climatic, edaphic, and vegetation gradients. Certain species (e.g., black bear) span the full range of elevation, the entire moisture gradient, and all vegetation successional stages. Other species such as the red tree vole apparently face strong "invisible barriers" (Grinnell 1941b), only occur in the canopy of a few conifer species (except under unusual circumstances), and are totally limited to western Oregon and northwestern California. Thus to some species an expansive clearcut and second-growth forest may appear as no barrier whatsoever, while another species could be totally marooned by the same conditions. Degree of isolation must therefore be viewed as a continuum that is species-specific and is dictated as much by the biology of the species as by environmental conditions. Findley and Anderson (1956) noted that arboreal species such as the squirrels, and other species such as marten and fisher that are highly dependent on forest habitat, were greatly limited by the Green River and its canyons. They also noted that the distribution of mammal species in the Colorado Rockies was inverse to their dependence on forest. The existence of riparian forest that could be used as a dispersal corridor was critical to the distribution of many species (Findley and Anderson 1956).

Yet an additional indirect effect of isolation is of special importance to animals. Because the dispersal of plants to isolated habitats is as problematic as is the dispersal of animals, isolation may also have the effect of reducing plant resource diversity and habitat quality (Johnson 1975). Thus, even if an animal species reaches a given island, the habitat might not be sufficiently diverse and complex to support it. Johnson (1975) concluded that both the impoverishment of bird species and the lower density of species that occurred on Great Basin mountains were due to the impoverishment of plant and insect groups upon which the birds depend for food and habitat. In support of this pattern, Hickman (1968) found that the average number ($\bar{x}$ = 34.1) of disjunct species occurring on peaks near ridges was over twice as great as the number ($\bar{x}$ = 15.1) occurring on isolated peaks.

The combined effects of all these factors seem to operate most significantly on amphibians and reptiles, followed by mammals, permanent resident birds, and then migratory birds. In other words, birds are proportionally more abundant in isolated communities such as Mary's Peak and the Steens Mountains than in the average Cascade fauna (fig. 6.12). Isolation is also less important in determining what type of bird species occur on montane islands than it is for determining mammals (Brown 1978).

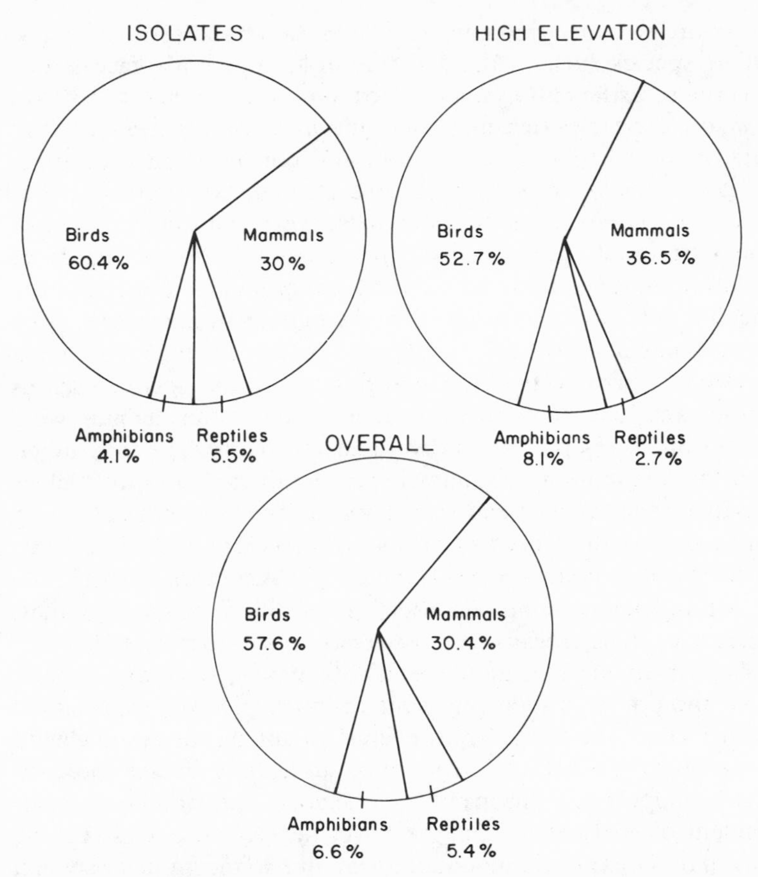

Figure 6.12 Percentage of terrestrial vertebrate species occurring in various taxa. Amphibians, reptiles, and mammals become relatively less abundant in isolated areas as birds become relatively more abundant (see also Brown 1978). Mammals become relatively more important than birds at higher elevations (see also Bratz 1952; Blake 1926).

The Distinction between True Island Biogeography and Habitat Islands

In spite of the great similarity between the biogeography of true oceanic islands and that of montane and habitat islands, several distinctions must be drawn (Wilcox 1980). True islands are surrounded by a medium that may function well as a dispersal agent but will not suffice as habitat for terrestrial species. Mesic mountain tops surrounded by a dramatically different environment such as desert are sufficiently analogous to warrant comparison. A clearcut totally surrounding an old-growth island is also sufficiently different from the old-growth habitat and sufficiently "hostile" to many old-growth inhabitants to justify using the analogy. But, as demonstrated by the earlier ordinations (figs. 5.10, 5.11, 5.12) and faunal similarity coefficients between old-growth and mature managed forest (table 8.1, p. 112), the analogy is not consistently strong. And herein lies a key to management strategy. To the degree that old-growth ecosystems can be surrounded, even partially, by similar habitat, the island analogy and its portents do not apply.

The second distinction between true islands and future old-growth islands may have more severe consequences. True islands are always thought of in reference to a nearby "continent" that serves as a source pool of immigrants. The "continent" is implicitly thought to contain a greater number, and indeed an inexhaustable supply, of potential colonizer species. Immigration and colonization from the source pool of species are always possible, no matter how improbable. However, some decades in the future when the regulated forest is composed of short-rotation monocluture plantations and when many species are restricted to widely separated preserves, what is to constitute the "continent"? There is no "continental" source pool of grizzlies to recolonize Mount Rainier National Park. Similarly, there will be no "continental" source pool of species for the replenishment of old-growth islands. As dramatically demonstrated for the eastern United States (Burgess and Sharpe 1981), it seems fair to presume that the old-growth islands will occur in a sea with no "continent." While the habitat patches will represent the islands, it will be the system of islands that must serve as the one and only species pool. There will be no larger reservoir of species from which immigrants can originate.

Fortunately, a regional system of national parks, wilderness areas, and research natural areas is in existence in the Cascades. The four parks are North Cascades (789 mi^2, 2,044 km^2), Mount Rainier (377 mi^2, 975 km^2), Olympic (1,400 mi^2, 3630 km^2), and Crater Lake (250 mi^2, 650 km^2). Of these, North Cascades is at the very northern end of the range and thus has very high latitude. Crater Lake is high altitude and has little old-growth Douglas fir. Olympic National Park is not in the Cascade Range and is rapidly becoming isolated by urbanization. The eleven wilderness areas average about 235 square miles (609 km^2) in size but the eight occurring in lower Washington and Oregon average only about 100

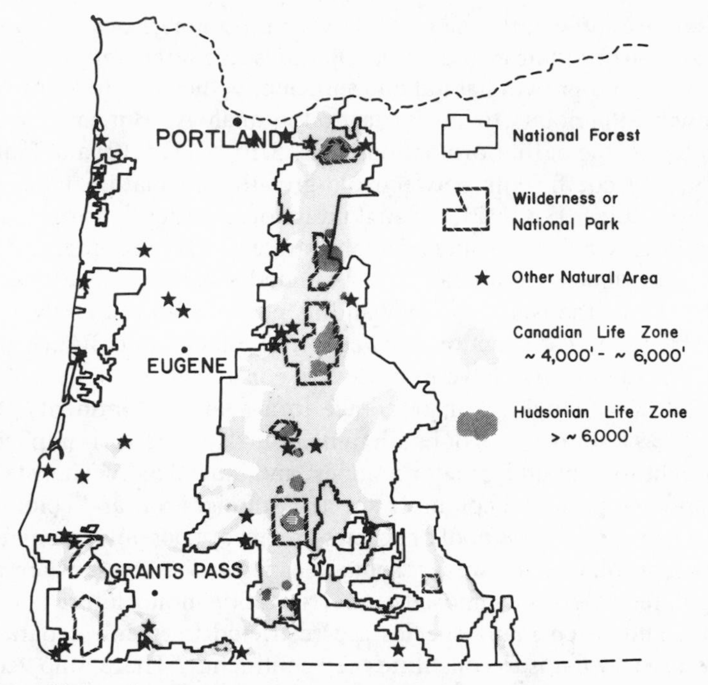

Figure 6.13 Location of western Oregon national forests, national parks, wilderness areas, and other forested natural areas against the backdrop of the high-elevation Canadian and Hudsonian life zones (Bailey 1936). The overwhelming majority of the protected area occurs at elevations above the ranges of thirty-five amphibian, reptile, and mammal species (figure from Harris et al. 1982).

square miles (270 km^2). The great disadvantage of these areas is that they generally occur at high elevations (fig. 6.13), and with few exceptions well above the zone of high species richness. Research natural areas (appendix 5) are well distributed throughout the region but they tend to be very small, with an average size of 430 acres (174 ha) in Washington (excluding Rattlesnake Hills) and 2,240 acres (908 ha) in Oregon.

Despite their inadequacies, these areas form an impressive arrangement of building blocks upon which a conservation strategy may be hinged. (fig. 6.14). Such a strategy will be proposed later in the text. It should be noted here, however, that preserving old

Figure 6.14 (opposite) The present system of national parks, national forests, wilderness areas, and other forested natural areas in the western Cascades (data and location map compiled by Sarah Greene and Bob Frenkel).

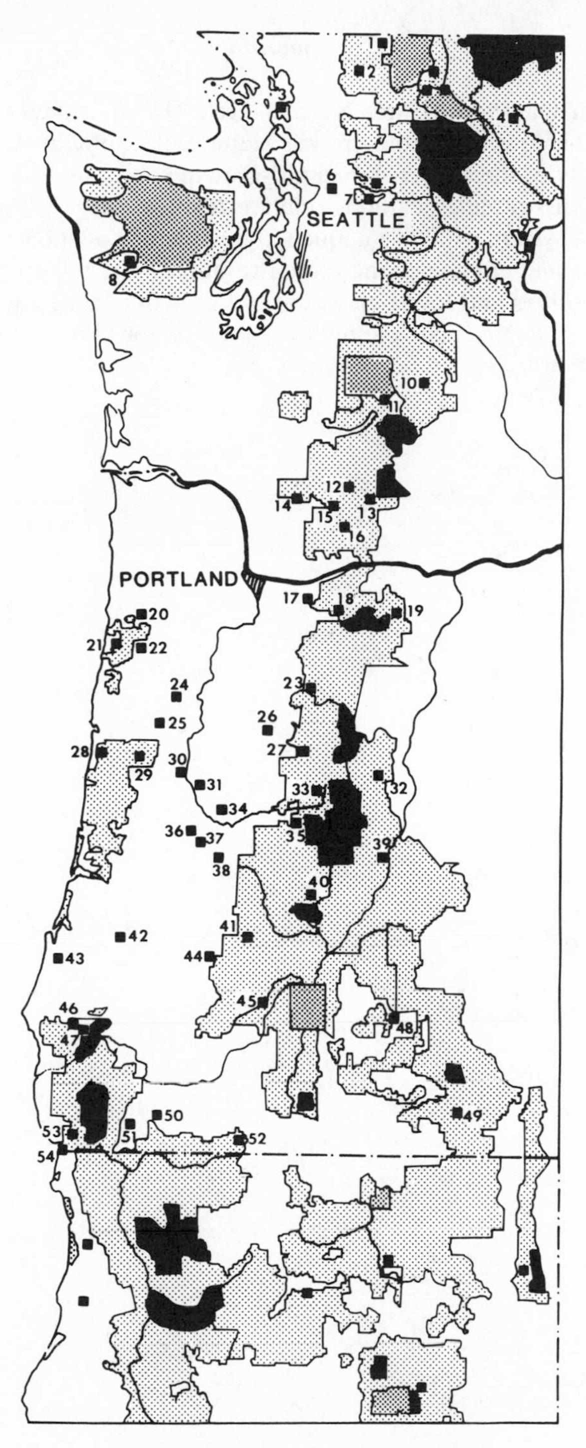
SEATTLE
PORTLAND
1
2
3
4
5
6
7
8
9
10
11
12
13
14
15
16
17
18
19
20
21
22
23
24
25
26
27
28
29
30
31
32
33
34
35
36
37
38
39
40
41
42
43
44
45
46
47
48
49
50
51
52
53
54

growth near the Research Natural Areas (RNAs) could not only be advantageous to the old-growth fauna, but also help maintain the integrity of the RNAs. When old-growth habitat islands are designated in strategic positions relative to these large preserves, the overall system will have a much higher chance of successfully conserving and enhancing the faunal resource. The integrated system of preserves and lower elevation old-growth islands must serve in place of the original "continent" of continuous boreal habitat and attendant species.

7

Genetic Resources and Biotic Diversity

Larry D. Harris, Michael E. McGlothlen, and Michael N. Manlove

Given a commitment to the maintenance of biotic diversity, the maintenance of viable populations, and to the preservation of representative old-growth Douglas fir ecosystems, the problem reduces to how much total acreage, what size tracts, and what distributional characteristics should be considered. In previous sections we have established the relevant states of nature that prevail and the applicability of insular biogeography principles. This chapter aims to clarify the issues surrounding the maintenance of diversity and will discuss the criteria for choosing one alternative approach over another. Discussion of diversity will be restricted to living biotic diversity and will not include such resources as edaphic, aesthetic, cultural, or other types of diversity. On the other hand, genetic diversity occurs at several hierarchical levels, only one of which is species diversity. Preservation of community attributes such as coevolved predator-prey relations and ecosystem processes such as nutrient cycles must be accomplished at higher levels of hierarchical organization.

Frankel (1974), Greig (1979), and others have established that conservation of wild biota is critically important, and that long-term conservation will only be successful in a climate of continuing evolution. Evolution, in turn, can only proceed adaptively if sufficient genetic variability is present to allow natural selective forces to discriminate between adaptive and maladaptive traits. This means that it is imprudent to wait until a species is nearly extinct before we begin to conserve it, as it will have already lost a significant portion of its genetic variation.

It is also important to appreciate that conservation of genetic resources is not so much aimed at protecting what occurs at present as it is at providing for the future. Strategies that do not allow the evolution of populations, species, and ecosystems are not ideal since they impair the flexibility of future generations (Frankel 1974;

Greig 1979). This implies that we should favor approaches that not only maintain biotic diversity but allow it to evolve and adapt to ever-changing conditions.

Endangered Species

One approach to the conservation of genetic resources is endangered species preservation. This approach holds that no known species should be forced to extinction by man's activities, whether benevolent or not. Once a species is classified as endangered, then conservation efforts should be directed toward it without reservation; aggressive, species-specific research and management should be implemented. Although this approach has been necessary on numerous occasions and has worked well, it is not ideal for several reasons. Since maximum gains can be made by focusing on specific aspects of biology and habitat, it frequently leads to a naively simple and often artificial approach to conservation: "Tell us the detailed habitat requirements so that we can address them specifically." In Florida this has led to inducing bald eagles to nest on fabricated platforms on artificial poles. This may be acceptable species management, but it is unacceptable genetic conservation. Perhaps worse, the species-by-species approach tends toward a never-ending marathon of salvage operations. Even when the species is secure there is no assurance that genetic diversity on other levels has been preserved. Allelic polymorphism or within-species variation is one aspect of diversity that is not ensured. Therefore, even though endangered species recovery is a necessary activity, it is not sufficient in and of itself.

Within-species Diversity

Just as genetic variability exists between species, variability within a species is also an important component of diversity. The proportion of loci in a population that are polymorphic (i.e., that possess more than one allele), the number and type of alleles at these polymorphic loci, and the average level of heterozygosity per individual in a population are three important measures of genetic variability. Average heterozygosity may be important to fitness as demonstrated by its positive correlation with reproductive success in white-tailed deer (Johns et al. 1979) and old-field mice (Smith et al. 1975). The variability present in a population is important for adaptive flexibility and future evolution. One important component of this variability is the presence of alleles that contribute to adaptability. However the preservation of aberrant or maladaptive genotypes (such as albinos in zoos) is not a positive contribution to variability or genetic diversity. The other main component of genetic variability depends on the extent to which the population is divided into subpopulations (lines) and the amount of inbreeding within these lines. In the absence of selection for specific traits, inbreeding increases the total genetic variance of the population (Allard 1960, 203). Variance within the lines increases in the early stages of inbreeding because of an increase in the frequency with which rare recessive alleles are expressed in homozygotes. As inbreeding continues, this within-line variance will decrease as some alleles are lost. Between-line variance also increases as subpopulations differentiate, and this increase more than offsets the within-line loss of variance. Thus the overall genetic variance is increased by dividing the population into inbreeding lines (Falconer 1981, 242–43; Robertson 1952). In the presence of uniform selection on all lines, this between-line variance would be much reduced. If selection operates differently on the different lines because of environmental factors, the increase in variance due to inbreeding may be even greater and lead to formation of discrete ecotypes.

There are several aspects of allelic polymorphism to be considered and several factors that affect these aspects. That organisms can exist without much local variability is demonstrated by plants that reproduce asexually and thus have entire demes with identical genetic makeup. But without allelic diversity, such demes are unlikely to outcompete closely related genotypes over a wide range of habitats or adapt to temporally changing environments. Such inflexibility is likely to lead to extinction. Because of physical, climatic, edaphic, and other gradients in nature, local adaptations based on genetic polymorphism are observed in many species. These variations can be manifested in a continuous gradation as a "cline" or as separate demes with little or no interchange of genetic material between them. When little genetic variation exists between individuals within a deme, then selection may operate at a higher level, and entire demes either survive or become extinct on the basis of their fitness and potential for adaptation.

In order to maintain adaptability, it is important to conserve as many genotypes of each species as possible. Much allelic diversity is overlooked because it is cryptic and yet its existence may be critical for future productivity gains (Campbell 1979; Silen and Mandel 1982) or population stability. The long-term viability of a species is founded on conservation of numerous demes or ecotypes throughout the geographic range of the species. Consistent with this is the preservation of local adaptation rather than allowing it to be indiscriminately destroyed by artificial mixing of genetic stocks through practices such as transplanting. "An ever-present danger . . . of introduction or re-introduction is the loss of the genetic distinction of a local endemic form" (Cowan 1965). The case against genetic contamination from translocation has been strongly made by Greig (1979). While translocation is frequently explained as an effort to help overcome inbreeding in small populations, there are many cases where it has done more harm than good by causing the loss of adaptation. Greig (1979) cites several examples of populations being extirpated because breeding seasons were disrupted by the introduction of animals adapted to a different photoperiod. In other cases, such large numbers of the exotic were introduced that the indigenous form was swamped and the resulting hybrid no longer exhibited the characteristics of the original population. This occasionally happens even without translocation by humans, as shown by the red wolf being swamped by the coyote (Shaw and Jordan 1977). The threat of inbreeding is not sufficient reason for translocation except when natural genetic interchange has been disrupted by alteration of the environment.

Inbreeding is the topic of greatest concern to geneticists when small populations are discussed (Ford 1964; Wright 1969; Lovejoy 1977; Miller 1979), and thus we will treat the topic in some detail. Our aversion to inbreeding derives in part from observations of the detrimental effects of marriage between close relatives. Disorders and defects occur more frequently in the progeny of related parents than in the population at large (Sutton 1966). But, since individual humans manifesting genetic defects are kept alive and are sometimes permitted to reproduce, the human situation is not directly analogous to that of natural populations of wildlife. Even in a random-mating wildlife population an individual will occasionally mate with a relative. As a population is reduced in size or divided into small demes, the frequency of inbreeding due to chance will increase. "Closed populations . . . undergo inbreeding at rates inversely proportional to population size" (Koger 1977, 434).

Inbreeding is a process whereby genes from a common ancestor are brought together by the mating of relatives. This tends to reduce genetic variation since the progeny may receive duplicate alleles from the common ancestor. The effects of this reduced variation may be positive, neutral, or negative. Because only a few positive effects (e.g., embryo compatibility and growth uniformity of crops) are known and the neutral outcomes are not observed, the negative effects receive most attention.

The negative effects of inbreeding can be observed at the level of the population but are most generally observed in individuals. Inbred populations have less potential for adaptation to environmental changes. The consequences of inbreeding observed in individuals can be placed in four categories: fetal survivorship, neonatal and juvenile survivorship, competitive ability, and adult fertility and fecundity.

Inbreeding depression refers to the decrease in fitness resulting from reduction in any of the four parameters named above, and has been repeatedly demonstrated with a wide variety of animals (Wright 1977; Soulé 1980). However, the extent to which it limits fitness and reproduction in a population varies with the type of organism and the circumstances. Increased fetal loss from consanguineous matings has been found in domestic, laboratory, and captive animal populations. In birds it is especially observable as a decreased hatching rate (Wright 1977; Soulé 1980). Comparison of inbred with non-inbred progeny of several species of captive wild ungulates reveals higher rates of juvenile mortality in the inbred animals (Ralls et al. 1979).

Lowered reproductive performance by inbred animals has been noted in domestic and laboratory animals (Wright 1977; Koger 1977). In small inbred populations, it is difficult to separate the effect of inbreeding on reproductive success and parenting ability from progeny viability. Decreased fertility of inbred animals has been demonstrated with domestic and laboratory animals, and the restoration of fertility upon outcrossing inbred lines also has been noted (Falconer 1981; Koger 1977; Wright 1977). This lowered fertility is of course correlated with decreased size and vigor of inbred animals. Inbred individuals that survive to become adults in a wild population may still be less successful in competing for mates. The combination of lowered viability and fertility can cause severe problems for small populations.

The magnitude of inbreeding depression depends upon factors such as mutational genetic load (hidden deleterious genes), intrinsic heterozygosity, the environment in which the population lives, and the reproductive capacity of the species. For example, a population living in a uniform environment may not suffer the effects of loss of adaptability to the same degree as a population occurring in a changing environment. Similarly, a rapidly increasing population with low-intensity selection will manifest more maladaptive traits than a population under intense selective pressure. Species that have a low reproductive potential will carry maladaptive traits much longer than species with high reproductive capacity. Thus a population that has a high rate of reproduction and that is subjected to intense selection will manifest comparatively few adverse effects of inbreeding when compared to one with a lower rate of reproduction and little selective pressure.

The cause of inbreeding depression can be either the loss of heterozygosity, the expression of deleterious recessive alleles, or a combination of the two (Franklin 1980). The mutational load (the genetic load of Muller 1950) of deleterious genes in a population causes a loss of fitness. The mutational load in most natural outbreeding populations is quite high. But since each individual is in competition with others that also carry a genetic load, it is possible for individuals to survive with much less than maximum fitness (Wallace 1970). When relatives mate, the deleterious genes from their common ancestor are more likely to be expressed in homozygous offspring.

Inbreeding coupled with selection can purge a population of its mutational load as the recessive deleterious alleles are expressed in homozygotes and then systematically eliminated. If the mutational load is overly high or if the rate of inbreeding is so great that selection cannot remove the alleles as quickly as they are expressed, then deleterious traits may become fixed in the population and the normal or advantageous alleles at this locus are lost. This can cause a loss of viability that cannot be recovered except by the rare event of a mutation restoring the advantageous form of the gene.

A population that has already undergone some inbreeding and has reduced its mutational load will not be affected by subsequent inbreeding to the same extent as a population that has not inbred at all and is carrying a high genetic load. Pere David's deer (Ralls et al. 1979) and the European bison (Slatis 1960) have not reflected the inbreeding depression for viability that has emerged in most other captive ungulates. Perhaps this is because most deleterious recessive alleles were already selected out during their long period at very low numbers. A comparison of attempts to establish inbred lines in different animals showed that Japanese quail failed to survive four generations of inbreeding while domestic fowl could survive fifteen (Soulé 1980). Again, the domestic fowl had probably been purged of most deleterious mutations while the quail were probably from an outbreeding population with a high mutational load. This would make them more vulnerable to inbreeding depression.

Experimental evidence to support this thesis has only recently appeared. Early experiments with mice carried out by Roberts (cited in Frankel and Soulé 1981, 68, and by Falconer 1960, 255) only showed that outcrossing inbred lines led to recovery of litter size, which had been lost by inbreeding. It has now been shown that such lines derived by outcrossing inbred lines are far more resistant to inbreeding depression than are outbred populations (Musiałek 1980; Lorenc 1980). Only four pairs of mice in outcrossed, previously inbred lines were enough to maintain fertility, while ten pairs were necessary in the outbred control population. This resistance to inbreeding depression is attributed to the previous elimination of the genetic load (Musiałek 1980).

Heterozygosis refers to the presence of different alleles at the same locus in an individual. Heterozygote advantage or hybrid vigor refers to the heterozygote's (Aa) being more fit than either homozygote (AA or aa). If this advantage refers to performance at a single function, it is called overdominance. It has been stated that little or no classical overdominance exists (Frankel and Soulé 1981, 44), meaning that the homozygotes will always function at least as well as the heterozygote. However, heterozygosity can definitely impart some advantage. For example, the presence of one allele may allow an individual to function better under certain conditions or during certain life stages while the second allele allows the organism to function better at different times or life stages (pleiotropy). Thus the heterozygote may be more fit when the entire life cycle is considered or when a highly dynamic environment exists. Consider a second example where the homozygote AA results in six litters of two individuals each and the homozygote aa results in two litters of six individuals each. Both homozygotes produce a total of twelve offspring, but with no overdominance at all (assuming only additivity), the sixteen offspring (four litters of four each) of the heterozygote Aa represents a 33% increase in total offspring over either homozygote.

Heterozygosis is obviously decreased as inbreeding increases homozygosis. When heterozygote advantage occurs, then inbreeding reduces it, in turn reducing fitness. Epistasis refers to the enhanced genetic effect resulting from complementary combinations of alleles at different loci. These combinations are less likely to occur as inbreeding decreases both heterozygosity and the number of alleles at relevant loci.

Some wild populations and most domestic ones routinely inbreed. This behavior will be selected for when mortality losses resulting from the process of outbreeding are greater than the effects of inbreeding depression (Bengtsson 1978). Smith (1979) reports that female fallow deer usually stay in the herd of their birth and frequently mate with their father. This is one method of increasing the number of genes like her own that the female passes on to the next generation, a form of altruistic behavior that results from kin selection. Smith calculates that it would take a one-third loss in fitness due to inbreeding to offset this selective advantage. The inbreeding depression of sire-daughter mating in most ungulates appears greater than one-third but the fallow deer has apparently reduced its mutational load to the point where this is advantageous behavior.

Since variability allows adaptation to changing environmental conditions, inbreeding is undesirable to the extent that it causes loss of variability. We emphasize, however, that given the complexity of and lack of knowledge about the mating systems and subtle adaptive history of different populations of many species, one must be cautious when assessing the actual or potential effects of inbreeding. Management decisions, for example restocking, should be based on more than just the desire to replenish local population numbers or the introduction of "new blood" into a population. The adaptive history of the source populations from which the individuals are transplanted must also be considered.

Bottleneck and founder effects occur when a population is temporarily reduced in number or when a new population is founded by only a few individuals. Under such circumstances genetic variability will be decreased, even if the population quickly increases to a viable level. In effect, a sample of the original population is taken; this sample may be representative or aberrant and it may contain most of the variation in the original population or very little of it. The percentage of original genetic variation retained by the reduced population is directly but geometrically related to the size of the bottleneck population ($\% = [1 - \frac{1}{2N}]\,100$). Thus, if a large population were suddenly reduced to ten individuals, they would be expected to exhibit 95% of the original variance; if reduced to fifty individuals they would be expected to contain 99% of original variability. While a large portion of total genetic variability may be maintained, bottlenecking will reduce the number of uncommon alleles at polymorphic loci (Denniston 1977). While this loss would not be expected to have much effect on fitness under current environmental conditions, it can limit adaptability to environmental change. Since inbreeding will further reduce this variability, and the inbreeding rate is itself a function of population size, the length of time the population remains at the reduced level is of vital concern. In other words, the size of the bottleneck is much less important than its length (Senner 1980). Therefore, a prolific species that can increase its numbers rapidly would not be expected to reflect as many consequences as species with a lower potential rate of population increases (Nei et al. 1975).

The best-known case of a large animal surviving a period of very small population size is the northern elephant seal. It was reduced to less than thirty prior to 1900 and has since recovered to more than 30,000. While almost no genetic variation is found in the present northern elephant seal population, the southern elephant seal, which did not go through such a severe bottleneck, shows considerable diversity (Bonnell and Selander 1974). While there is no definitive explanation for why the northern elephant seal recovered so well, it should be noted that its maritime habitat does not require adaptation to a wide variety of environmental types. Since there is so little variability remaining, adaptability may now be impaired and the population may be vulnerable to future environmental perturbations. At the present large population, considerable genetic variability may be restored by mutation over a few generations.

Minimum population size is a critical consideration for several reasons (Shaffer 1981). The probability of extinction from random population oscillations is of concern, as are social requirements for some species. Darling (1952) suggested that the heath hen and the passenger pigeon may both have become extinct because their numbers dwindled to such an extent that they did not initiate breeding. Other social functions such as migratory behavior may require some minimum number of individuals. While the factors described above determine the minimum level at which the population maintains its ability to recover, the minimum population size needed to prevent adverse effects of genetic drift is probably higher, and the size necessary to ensure genetic flexibility may be higher still. This is the threshold that must be considered when choosing a conservation strategy.

Random genetic drift is caused by the sampling of parental genes in meiosis and the random combination of these genes in the next generation. In a very large population no change in gene frequencies would be expected as a result of random effects, but in a very small population there can be significant changes each generation (Crow and Kimura 1970; Kimura and Ohta 1971). The change in the genetic composition of a population that results from these random effects is known as genetic drift. Selection, particularly stabilizing selection, can help to neutralize the effects of genetic drift for intermediate gene frequencies, since alleles tend to be either fixed or lost by genetic drift (Hartl 1980; Falconer 1981). Selection for (and adaptation to) local conditions or a changed environment can be mistaken for effects of random drift, and this makes it difficult to estimate how significant drift is in wild populations. Its effects are well known in laboratory populations, however, and it is found that many alleles are either fixed or lost in small populations. Once this end point is reached, variability can be restored only by mutation or immigration. Most disadvantageous alleles are rare and will therefore be lost quickly. Occasionally an advantageous allele will be lost and a corresponding disadvantageous one fixed. The smaller the population, the more often deleterious genes can be fixed. Since the loss of genetic variability and fitness can only be offset by mutation or immigration, the minimum population size is determined by the balance between the rate of drift and the rate of mutation and immigration that introduce favorable or neutral alleles.

Because different species manifest different age structures and breeding habits, the number of genetically active individuals might be different in two species of equal absolute size. In order to compensate for these discrepancies, the concept of effective population size (N_e) is used. Assuming equal numbers of breeding males and females, random mating, nonoverlapping generations, and normal diploid meiosis, the effective population size (N_e) equals the actual population size. Deviations from this idealized population make the effective population size substantially lower than the actual population. Since nonreproductive animals are not included in the tally of effective population size, young animals as well as nonbreeding adults must be subtracted. Another correction is demonstrated by the following equation, which compensates for unequal sex ratio in the breeding population:

$$N_e = \frac{4 \cdot n_m \cdot n_f}{N}$$

where n = number of a specific gender
and N = total number of individuals

It will be quickly noted that when the number of breeding males equals the number of breeding females, the actual breeding population equals the effective population (fig. 7.1). On the other hand, if fifty times as many of one sex participated in breeding as the other sex (e.g., harem animals), then the effective population size would be only 8% of the actual size (fig. 7.1). In the extreme case of only one breeding male, the effective population size will approach four even if the number of females approaches infinity (Wright 1977). Corrections for fluctuating population size, variance in number of progeny, and other factors are also possible (Franklin 1980).

Given that we can use effective population size to compare one species to another and generalize across species, we can calculate numerical estimates of minimum population sizes. For a very few generations a loss of 1% of the genetic variability per generation can be tolerated (Denniston 1977). Without dominance or epistatic variance, additive variance is lost at the rate of ½ N_e per generation. Therefore, if we set ½ N_e to 1%, it follows that N_e must not fall below fifty. Therefore, fifty is commonly cited as the short-term

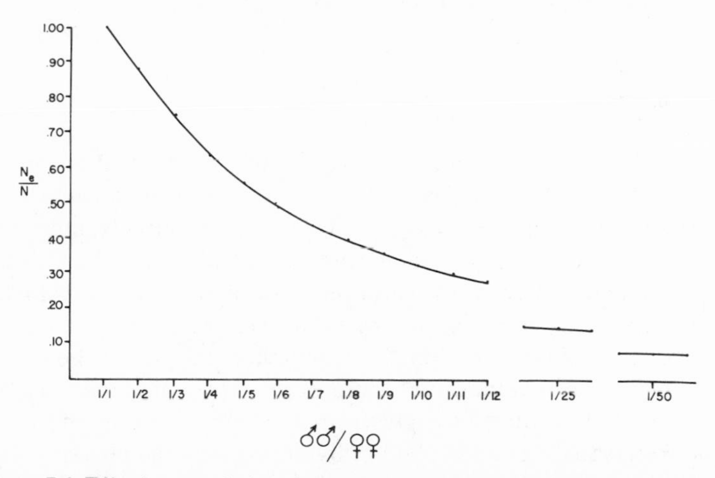

Figure 7.1 Effective population size as a proportion of actual population size in response to the effective sex ratio of breeding individuals. The effective population size of monogamous species with a 1:1 ratio of breeders is equal to the actual size. As the ratio of males to females is skewed, as in harem animals, the effective gene pool size is reduced to a small percentage of actual size.

emergency level that populations should not fall below (Soulé 1980).

Mutation cannot be expected to compensate for a loss of genetic variation exceeding 0.1% per generation (Soulé 1980). Therefore, an effective population size of 500 represents a minimum level for long-term maintenance of the population. We hasten to reiterate that the actual number of live animals would have to greatly exceed this figure in order to generate the N_e of 500.

Local population sizes can vary greatly within a species. Factors such as disease, predation, competition, and harsh weather can drastically reduce effective numbers in local populations to the extent that genetic drift can then have a significant effect on gene frequencies. Therefore, populations at the edge of their species' range or otherwise subjected to suboptimal environmental conditions may require special attention when management decisions are directed to the species as a whole.

Certainly we must consider the minimum effective population size for species of concern when making management decisions or defining management areas. Paradoxically, at the same time that we are introducing major changes into the environment (roads, pesticides, acid rain, etc.) and thus wildlife populations will require maximum adaptability, we threaten to reduce their ability to respond by insularizing populations, forcing them to inbreed, or exposing them unnecessarily to vagaries of genetic drift.

Faunal Preservation vs. Maximum Species Richness

On March 22, 1983, the *New York Times* published an article entitled "Clear-Cutting Exonerated." Referring to an article in *Science*, the conclusion was that clearcutting actually increased species diversity over the uncut control plots. The results are misleading and deceptive for several reasons, one of which is relevant here. Since no species lists are presented in either article, it is not possible to evaluate whether the researchers measured the occurrence of flora or vegetation. Flora refers to the list of species that are expected to be found in a given area. Vegetation, on the other hand, refers to the mass of plant individuals occurring on a site. This vegetation may bear little or no similarity to the flora. The objective of preserving biotic diversity refers to maintaining the flora, not the vegetation. High species diversity consisting of alien, ruderal, and/

or very common species is not an acceptable goal, especially if it occurs at the expense of the natural flora.

Fauna refers to the taxonomic list of animals that normally occur in an area. It is analogous to flora as described above and is to be distinguished from wildlife or animals, terms that refer to the mass of animal individuals actually occurring on a site. As detailed by Buechner for the state of Washington (1953) and described earlier in this text, naturalization of six European bird species and four additional North American bird species not native to the state adds a significant component to the avian community. Indeed, management for or against these species can be justified on many grounds. It should not be argued, however, that the faunal diversity has been preserved simply because the species richness is greater now than it was in 1900.

There are several obvious approaches to the maintenance of biotic diversity. One strategy would be to choose sites that support the highest species richness. In the case of terrestrial vertebrates in the Cascades, these would be low elevation, old-growth ecosystems on moist sites (Figs. 5.5, 5.6, 5.9, 5.11, 5.12). However, this strategy does not explicitly address the issue of endemic species, those that occur only in the specific locality, nor does it address the presently threatened and endangered species. Hickman (1976) has observed that 85% of the plant species occurring in the western Cascades do not occur in the forests dominated by Douglas fir, but rather are found on scattered nonforested tracts occupying only about 5% of the total area. Habitats most likely to contain threatened and endangered plant species are sphagnum bogs, sedge meadows, rock outcrops, ridge systems, spruce bogs, alpine rock gardens, and subalpine forest/meadow transition (Schaaf 1979). Fortunately, none of these sites has a high site potential for timber and thus can easily be spared the effects of intensive forestry operations. Therefore, emphasis will be maintained on the old-growth Douglas fir community type throughout the rest of this work.

If a considerable number of sites is to be included in the overall system of old-growth islands, then abundant opportunity will exist to accommodate different objectives. A majority of the sites may be chosen on the basis of their representativeness, while another subset may be chosen on the basis of their contribution to the overall system (e.g., "stepping-stone" islands). A third subset of islands may be chosen on the basis of their unique individual characteristics and the presence of endemic species. Rather than viewing these as competitive objectives, they should be seen as complementary approaches toward conservation of the entire genetic resource at the level of the landscape and region. By choosing many old-growth islands, the concept of island replicates (as distinct from duplicates) that incorporate the broad spectrum of biotic diversity will be embraced.

The Equivalence of Species

As demonstrated earlier, substantially different assemblages of species are associated with different successional stages. It was also pointed out that somewhat different assemblages are associated with different-sized habitat patches. Robbins (1979) and Anderson and Robbins (1981) have described this phenomenon in terms of "area-sensitive" species. Pursuit of the objective of maximum species diversity or even maximum species richness could lead to serious negative consequences if taken literally and if no regard were paid to the considerations outlined above. Without asserting that a given species is "worth" any more than another species, it can easily be demonstrated that some species are more easily managed for than others. Species such as English sparrows, domestic pigeons, robins, deer mice, rats, and oppossums occur commonly in human-dominated environments and hardly require timber management concessions in their behalf. Carnivores such as cougar, lynx, marten, fisher, and wolverine do not occur commonly in human-dominated landscapes, and therefore require special consideration. Kendeigh's (1944) concept of "interior" versus "exterior" species and the more recent concept of area-sensitive species address the same issue. Some avian (and perhaps mammal) species simply do not occur in small fragmentary patches of habitat that are dominated by edge characteristics and edge species. In this management-oriented context, all species should not be considered equally. Maintenance of diversity that derives from the substitution of species that are common for those that are rare should not be considered an acceptable alternative. Moreover, simple measures of species diversity that do not take this into account should be rejected as inadequate.

Ecosystems and the Landscape Mosaic

The approaches to conserving genetic diversity discussed above are all oriented toward species and components of diversity that have been identified and are known. However, there are no doubt species as yet undiscovered and many important ecological phenomena and processes that we know little about. We know little about the long-term role of terrestrial predators in community structure and ecosystem function or the relations between vertebrates, mycorrhizal fungi, and tree growth. As Cowan (1965) observed, "lost segments of the ecosystem take with them their unexposed truths." Gilbert and Raven (1975) and especially L. E. Gilbert (1980) present the essential argument for conserving entire biological communities, and it seems that the only level of hierarchy that is both necessary and sufficient to meet all objectives is the ecosystem or some higher-level approach. The strategy selected should not only ensure the conservation of spotted owls, but all the intricate linkages that are associated with natural populations of spotted owls in naturally functioning ecosystems. Many of these are as yet unknown.

As will be shown later, many of the larger vertebrates of the western Cascades have home-range sizes of thousands of acres (e.g., cougar, black bear) and cannot be conserved or managed *within* any single forest stand or small set of forest stands. These animals (the larger carnivores) are the product of, and most likely dependent upon, the landscape mosaic. Smith (1972), Crowley (1978), Harris and Smith (1978), Miller (1978), Pickett and Thompson (1978), Whittaker and Goodman (1979), and others have attempted to describe relations between the landscape mosaic and ecosystem function. Although most of the relations are poorly understood and inadequately demonstrated, it seems that a landscape mosaic approach is the only one that can ensure the conservation of (1) known endangered species; (2) ecotypic diversity and allelic polymorphism; (3) the full gamut of native vertebrates; (4) natural animal community interactions of native large mammals and birds; (5) unknown species and processes; and (6) known natural old-growth ecosystem processes.

8

Evaluation of Alternative Approaches

Previous sections of this study have summarized background information and presented selected principles of island biogeography and their implications. A review of the most obvious alternatives will assist the decision maker in moving from ecological implications to forest scheduling applications. The key variables involved in the analysis of alternatives are total area commitment, number of habitat islands, size of habitat islands, distance between and connectivity of islands, and location or choice of habitat islands.

The Total Area Requirement

Public debate regarding societal, and thus agency, commitment to old-growth management frequently centers on "how much is enough." The choice criteria and trade-off relations between different alternatives are rarely stated or known, but even if they were, the question would probably remain intractable. "How much is enough" is clouded by so many qualifications and value judgments that it remains a policy decision. A few of the considerations are given below.

Although old growth appears to provide an optimal habitat for many species of vertebrates, to date no species has been shown to be an obligate inhabitant of old growth. This means that the old-growth ecosystems may best be thought of as diversity islands and that other more subjective criteria must be considered. Since the largest native mammalian carnivore (grizzly) and the largest raptor (California condor) are already locally extinct, some may argue against attempted maintenance of the full complement of native species. Given that two species have already been excluded from consideration, why shouldn't two additional species (e.g., gray wolf and wolverine) be allowed to become locally extinct? It might also be asserted that the preservation of large carnivores that are not obligatorily linked to old growth should be separated from the old-growth issue and dealt with separately. This would be folly. There is no old-growth issue save for the desire and objective of maintaining biotic diversity. Moreover, I believe a single strategy aimed at maintaining the large, wide-ranging carnivores, maintaining biotic diversity, and maintaining representative old-growth ecosystems can be devised.

As demonstrated in figures 5.10, 5.11, and 5.13, there is a high degree of overlap in the species that use old growth as primary habitat and those that use mature managed stands of timber for primary habitat. A surrounding stand of mature managed timber could therefore represent a true multiple-use function by serving as support habitat for old-growth denizens while allowing responsible and economically meaningful harvest of the managed stands. A surrounding stand of mature timber will also buffer the old-growth ecosystem from climatic effects, significantly reduce the threat of fire, mitigate the negative effects associated with dramatic, induced edges, and reduce the novelty status associated with the occurrence of a distinctly different ecosystem amid the short-rotation plantations. On the other hand, surrounding expanses of clearcut would expose the old-growth island to maximum climatic impact, create maximum habitat differences for wildlife, facilitate and enhance the prospect for negative edge effects (see p. 131), dramatize the difference between ecosystem types and approaches to forest management, and force the applicability of the true island biogeography analogy. Thus the question of how much is enough can only be fairly addressed in the context of surrounding forest conditions.

Considerations of total area commitment involve only biological factors and say nothing of regional socioeconomic conditions. These further complicate the issue and reinforce the conclusion that the total area commitment to old-growth preservation and/or management is at present a policy question for which there are no clear biological answers.

Effective Habitat Island Size

Three factors that determine the effective size of an old-growth habitat island are (1) actual size; (2) distance from a similar old-growth island; and (3) degree of habitat difference of the intervening matrix. Obviously, if there are only subtle differences between the old-growth ecosystem and the structure of the intervening forest, then the old-growth stands may not appear as islands at all. The entire acreage might then be considered as a composite, effectively increasing the size of the old-growth island. This concept led Harris and his coauthors (1982) to conclude that in order to achieve the same effective island size a stand of old-growth habitat that is surrounded by clearcut and regeneration stands should be perhaps ten times as large as an old-growth habitat island surrounded by a buffer zone of mature timber (fig. 8.1). There are several reasons for this.

Fritschen and his associates (1971, 2) observed that "when the wind was blowing from the clearcut into the forest (upslope and up valley wind) the wind speed was reduced to an equilibrium state within 60 meters of the forest wall or 2 to 3 tree heights." This and other studies have led to the "three-tree-height" rule of thumb for how far climatic effects of a surrounding clearcut will penetrate into

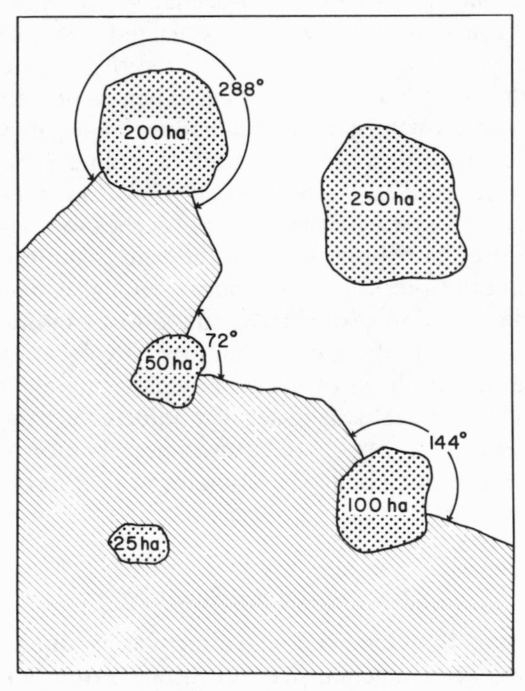

Figure 8.1 Inverse relation between recommended size of old-growth stands and the degree of insularity determined by surroundings. Old-growth stands are shown by dotted pattern, second-growth forests are shown by striped pattern, and the remaining area represents clearcut areas (figure from Harris et al. 1982).

an old-growth stand (Leo Fritschen, pers. comm.). Since area measurements have little meaning without the shape being specified, it is better to think in terms of "threshold thickness." The threshold thickness at which a central point would be buffered from winds from either direction would be on the order of six tree heights or 1,200 feet (366 m). A stand 1,200 feet in diameter (≃ 26 acres, 11 ha) would presumably have a central point where the wind speed had come to equilibrium. This point would have no area associated with it, however, and so any old-growth island expected to maintain its ecosystem integrity (considering wind, temperature, light, and relative humidity) would need to be of specified size plus the 600-foot (183 m) buffer. For example, a 200-acre (80 ha) circular old-growth stand would consist of nearly 75% buffer area and only 25% equilibrium area. A 775-acre (315 ha) circular stand would consist of 35% buffer area, and a 3,100-acre (1,250 ha) stand would consist of about 15% buffer area. A circular stand would need to be about 7,000 acres (2,850 ha) in order to reduce the 600-foot buffer strip to 10% of the total area. It is important to note, however, that the surrounding buffer stand does not have to be old growth, but only tall enough and dense enough to prevent wind and light from entering below the canopy of the old-growth stand. This means that mature planted forest could serve this function.

In reviewing the composition of bird census results, Kendeigh (1944) noted that patches of eastern deciduous forest greater than 65 acres (26 ha) had acceptably low percentages of "forest edge" species, and thus warranted consideration as valid avifaunal census plots. By inference, he concluded that 65 acres was of sufficient size to buffer the effect of edge species on forest interior bird species. Ranney et al. (1981) concluded that tree species composition would be affected in an edge zone approximately 170 feet (50 m) wide. Brittingham and Temple (1983) recorded negative effects of cowbirds parasitizing breeding bird nests as far as 1,000 feet in from the edge. Thus, the area requirements given above for the moderation of climatic effects may represent only a starting point for initial planning purposes.

A second way that surrounding mature stands increase the effective island size of the old-growth patch is through habitat similarity. If two habitat types supported exactly the same species of animals, the similarity coefficient for the two stands would be 1.0. If, on the other hand, the two habitats supported no species in common, the similarity coefficient would be 0.0. Amphibian, reptile, and mammal species occurrence in stands of six different successional stages was presented in figures 5.10 and 5.11. Those same data were used to calculate similarity coefficients of the animal communities inhabiting the six successional stages (table 8.1). Stage 5 (large sawtimber) is 97% similar to old growth and stage 4 (sawtimber) is 70% similar. On the other hand, the fauna of stage 1 (regeneration) is only 42% similar to that of old growth, and stage 2 (seedling-sapling) is only 46% similar. A surrounding buffer zone of large sawtimber would therefore seem to be the ideal matrix within which to manage old growth.

Other factors such as fire are also involved in long-term patch

TABLE 8.1

A. Number of amphibian, reptile, and mammal species that any two designated successional stages support in common. The upper right half of the matrix considers only the species using the stages as primary habitat, while the lower left half of the matrix considers all species that use the stage for either primary or secondary habitat. Numbers on the diagonal represent the number of species that occur in that successional stage.

B. Similarity coefficients between designated pairs of successional stages with the upper right half being based only on primary habitat; those in the lower left are based on both primary and secondary habitat.

$$S = \frac{2C}{A+B} \times 100$$

where: A = no. spp. in stage A
B = no. spp. in stage B
C = no. spp. in common

A.	1	2	3	4	5	6
1	98	79	34	34	34	34
2	94	94	35	35	34	34
3	79	79	86	36	35	35
4	78	78	85	88	38	38
5	77	77	84	87	93	65
6	77	77	84	87	93	93

B.	1	2	3	4	5	6
1	1.0	91	53	52	43	42
2	98	1.0	60	59	47	46
3	86	88	1.0	96	69	67
4	84	86	98	1.0	73	70
5	81	82	94	96	1.0	97
6	81	82	94	96	1.0	1.0

dynamics (e.g., Crowley 1978; Pickett and Thompson 1978; Veblen 1982), and these need to be considered. For example, Isaac (1940) observed that "broad walls of green timber between small openings reduce fire hazard and make necessary burning easier." He also observed that "the weed-brush stage is most subject to fire. . . ." It appears from these and other lines of evidence and field observations that the use of mature timber as a surrounding buffer is an expedient way to increase effective island size. If it is true that surrounding buffer zones of large sawtimber wil effectively increase the size of old-growth islands, this will have major implications for scheduling decisions.

Size vs. Number

For any given policy commitment to old-growth conservation there is a trade-off between size and number of stands. The choice can be a large number of smaller stands, a small number of larger stands, or any number of combinations in between. The tradeoff relation is not linear, but geometric (fig. 8.2). The three parameters—total area commitment, number of old-growth stands, and average size of stands—determine the decision arena. Any combination of the three parameters that does not violate one of the policy or algebraic constraints is a feasible alternative.

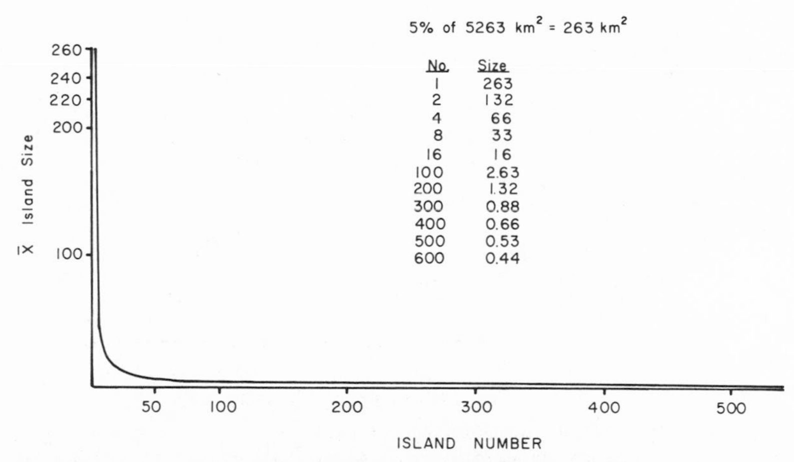

Figure 8.2 Trade-off relation between island number and average island size for the Willamette National Forest assuming a 20% commitment to 320-year long-rotation management islands which maintains 5% in old growth (>240 yrs) in perpetuity.

An old-growth island of one acre (0.4 ha) is surely not a worthwhile or feasible consideration. An old-growth island of ten acres (4 ha) is probably also an unrealistic alternative. Given that we have some guidelines for determining what is a minimum viable area, it is possible to add a further constraint which asserts that acceptable old-growth stands will be above some specified size. This can be entered into consideration as a "minimum size constraint." It further restricts the set of realistic or feasible alternatives (fig. 8.3). The same rationale can be used when considering the ideal number of habitat islands. In order to maintain the diversity of genetic ecotypes and to prevent catastrophic accidents from destroying the

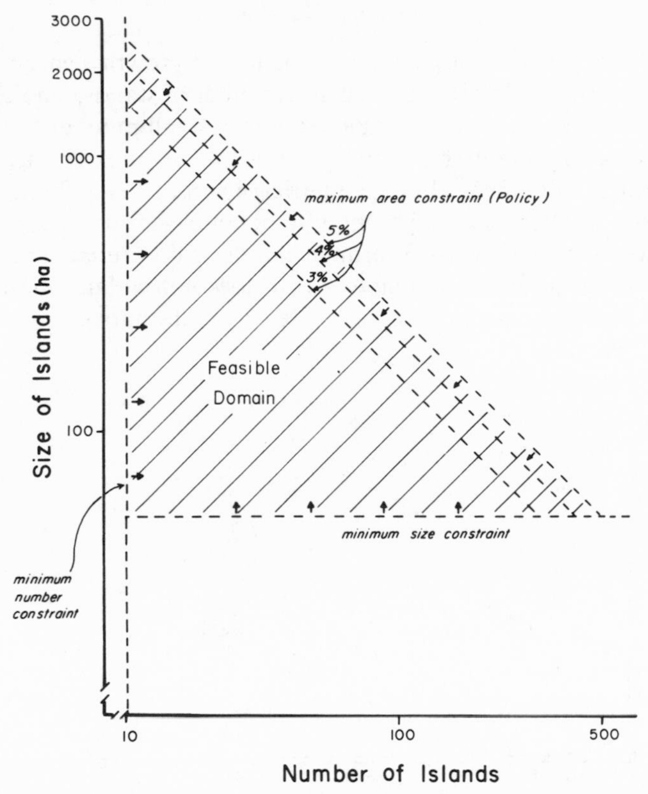

Figure 8.3 Trade-off relation between number of islands (log scale), average size of islands (log scale), and different policy constraints. A commitment of 20% long-rotation management would ensure 5% in old growth, a 16% commitment would ensure 4% in old growth, and a commitment of 12% would ensure 3% in old growth in perpetuity.

islands (e.g., Veblen 1982), more than one habitat island would be ideal. Whatever number is considered (e.g., ten) can be entered as a constraint (i.e., the minimum number constraint). This further restricts the feasible number of alternatives. The composite set of all remaining feasible solutions is referred to as the "feasible domain" (fig. 8.3).

Since the constraint lines are all variables, the total set of feasible solutions (i.e., the feasible domain) can be either small because of rigorous constraints or large because of relaxed constraints.

There are many other constraints that might be added to any formal problem definition. For example, there is not an unlimited number of old-growth habitat islands available. As noted in figure 4.6, a total of 319 stands of old growth remain in the Siuslaw National Forest. Thus for that forest, the set of feasible solutions cannot involve any combination of size and number that exceeds 319 islands. Similarly, the largest existing old-growth island is 970 acres (393 ha), and so all feasible solutions must involve old-growth islands this size or smaller. When realities such as these are imposed, the set of feasible solutions becomes very much more restricted (fig. 8.4).

When the problem is formalized as a trade-off relation such as this, the optimal solution is known to occur at the intersection of two or more constraint lines. In the graphical representations (figs. 8.3, 8.4) this will be represented by a corner point of the feasible domain (assuming the solution is not degenerate). The determination of which corner point, or the intersection of which two constraints, depends upon the trade-off relation between the decision variables, number of islands, and size of islands. If increasing the average size of old-growth islands is more effective in meeting the objective (i.e., if the slope of the objective function is slight), then the optimal combination will occur at one of the upper corners of the graph and involve the largest possible average-sized islands. If the objective is met more effectively by increasing the number of islands then the optimal solution will occur at one of the corners on the right-hand side and will involve the maximum number of islands. The optimal solution point is totally dictated by the value contributed (to the objective function) by island number relative to the value contributed by island size, and the specified constraints. Considerable research is necessary to determine the importance of size relative to number, but for now several additional factors can help guide decision makers.

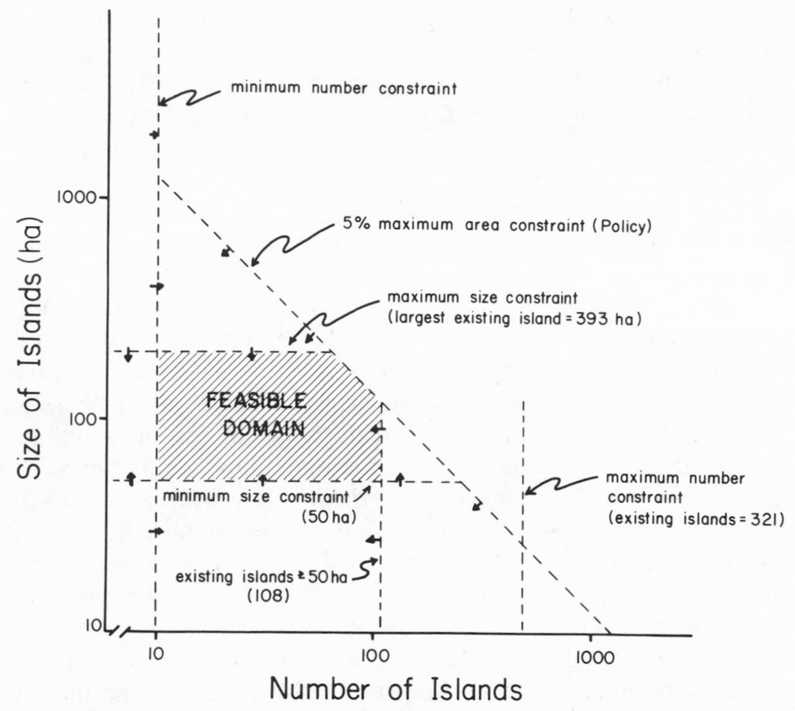

Figure 8.4 Decision framework for the Siuslaw National Forest in terms of various constraints. Assuming a commitment of 20% of forest acreage to long-rotation management, then 5% could be maintained in old-growth islands in perpetuity. If substantially surrounded by mature timber, then perhaps 125 acres (50 ha) would suffice as a minimum size constraint.

Interisland Distance

It was pointed out earlier that habitat islands were fundamentally different from true islands in that the future set of habitat islands will probably contain the total species source pool. This means that interisland distance may be a much more relevant parameter to maintenance of viable populations than is the case with true islands that draw species from a continental source pool.

The number of potential interactions between elements in a population is a strong power function of the number of elements involved. For example, when the number of elements is two, there is only one interaction; when three, there are three two-way interactions; when four elements, then six pairwise interactions, and

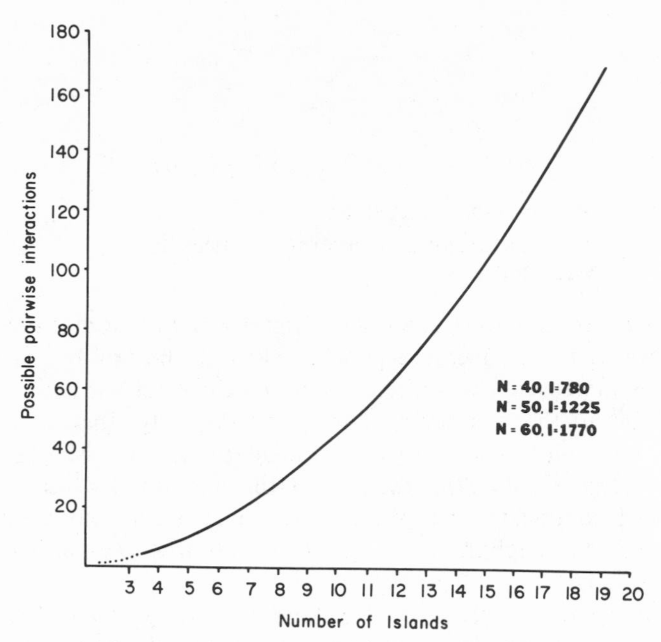

Figure 8.5 Geometric relation between the number of possible pairwise interactions between habitat islands and island number. In a two-dimensional forest environment not all of these possible interactions are equally probable or meaningful.

so on. When the number of elements is increased to ten, the potential two-way interactions are forty-five, and when $N = 20$, then $I = 190$. The number of potential island interactions increases very much faster than the number of islands (fig. 8.5). When limited to two-dimensional space in a tract of forest, not all pairwise interactions have equal probability of use. Still, this formulation serves to illustrate the magnitude of the relation between the number of islands and the number of potential interactions. If interactions between islands are important, then it should be noted that slight increases in island number imply geometric increases in the potential interactions.

Estimation of the average distance between an element in a population and its nearest neighbor has been useful in forest mensuration and plant ecology (Evans and Clark 1954). If the elements are distributed randomly then the average nearest-neighbor distance is given by the formula:

$$D = \frac{1}{2\sqrt{n/A}}$$

$$D = 0.5\,(A/n)^{0.5}$$

$$\log D = \log 0.5 + 0.5 \log (A/n)$$

where: D = average nearest neighbor distance
n = total number of elements (islands)
A = total area

It will be seen from this equation and figure 8.6 that average nearest neighbor distances decrease geometrically as the number of elements is increased. Again, if a small nearest-neighbor distance is important to the maintenance of biotic diversity, then a larger number of islands is far superior to a smaller number of islands. It is also of interest to note that the form of this equation for interisland distance is identical to that of the species-area relation described in figure 6.3. Although there are implications to this, they will not be developed here.

An alternative analysis of interisland distance is also instructive. If islands could be chosen to occur systematically on square grid points throughout the forest area, then each island would have four neighbors at equal distances. This is a way of maximizing average nearest-neighbor distance with the qualification that the number of equally nearest neighbors is four. Although actual distances are different, it will be seen that the pattern of nearest-neighbor distance between systematically placed islands is generally the same as for random placement (fig. 8.6). Many different distribution patterns will fall somewhere between the randomly distributed and systematically distributed patterns described. These two can therefore be used as limits within which interisland distances can be judged.

A specific example of trade-off relations between island number, average island size, and interisland distance for a specified area commitment is given in table 8.2. It should be recalled that (1) the number of possible islands decreases geometrically as the average island size increases (fig. 8.2); (2) the average nearest-neighbor distance decreases geometrically as the number of islands is increased; and (3) the number of possible two-way interactions increases geometrically as the number of islands increases.

The importance of different island sizes and interisland distances cannot be fully appreciated unless evaluated with regard to the

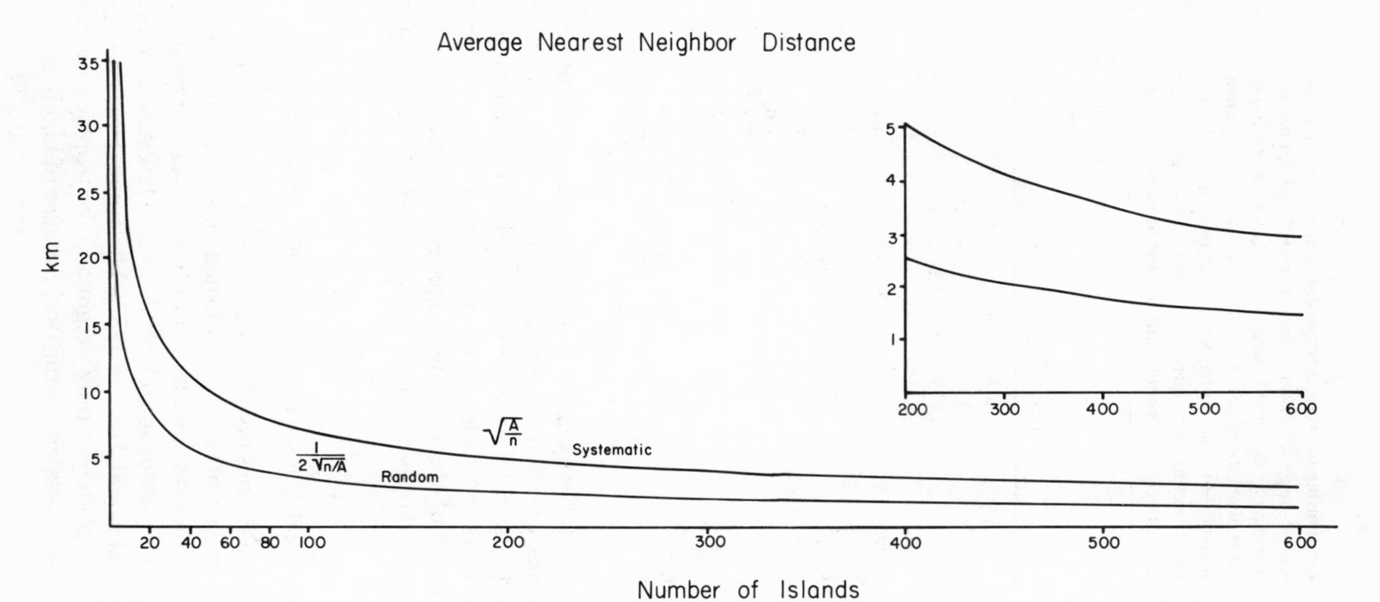

Figure 8.6 The average nearest-neighbor distance between habitat islands increases geometrically as the number of islands is reduced.

TABLE 8.2
Relationship between average size, number, and average nearest-neighbor distance for old-growth habitat islands assuming a 5% commitment (65,000 acres, 26,300 ha) of the Willamette National Forest. The random placement model is based on center points with no individual area. The systematic placement model is based on edge-to-edge measures to four equidistant nearest neighbor islands.

		Average nearest neighbor (km)	
No.	$\bar{X}$ size (ha)	systematic	random
1	26,300		
2	13,150	39.8	25.6
4	6,575	28.2	18.1
16	1,644	14.1	9.1
50	526	8.0	5.1
100	263	5.6	3.6
200	132	4.0	2.6
400	66	2.8	1.8
526	50	2.4	1.6
700	38	2.1	1.4
800	33	2.0	1.3
900	29	1.9	1.2
1,000	26	1.8	1.2

species involved. For example, it does little good to calculate potential island sizes without reference to the species that might be contained by the habitat patches. Little significance can attach to interisland distances unless they are measured against probable travel distances of specific animal species. Estimates of terrestrial mammal home ranges are given below to provide such a reference point.

Mammal Home-range Sizes and Travel Distances

Whether the species involved are fossorial, cursorial, aboreal, or of some other type, it is well established that mammal home-range size is predictable on the basis of body weight, percentage of flesh in the diet, and mode of life (McNab 1963; Harestad and Bunnell 1979; Gittleman and Harvey 1982; Mace and Harvey 1983). A simplified approach uses only weight and the trophic categories of herbivore, omnivore, and carnivore. Because of highly inadequate site-specific data on food habits and weights, the following approximation equations are used. When weight is given in grams, the home range in hectares is given by the least-squares prediction equations (equations from data in Harestad and Bunnell 1979):

$$H = 0.002W^{1.02} \text{ herbivores}$$
$$O = 0.059W^{0.92} \text{ omnivores}$$
$$C = 0.022W^{1.30} \text{ carnivores}$$

The body weights used for calculations are 80% of the maximum weights reported by Maser and his coauthors (1981).

Three major uses can be made of the results (table 6.3). If the conservation strategy is aimed at choosing old-growth patches sufficiently large to "contain" the animals within, the calculated home-range sizes will be useful in estimating the patch size necessary for individuals of the various species. If old-growth habitat islands were being established with a particular species in mind, then the area of the habitat islands could be compared with the home-range projection to assess adequacy. It should be noted that thirty-six of the fifty-eight species considered have projected home-range sizes of less than twenty acres (8 ha) (table 6.3, fig. 8.7). On the other hand, projected size for the ten species with the largest home ranges is greater than 2,000 acres (940 ha).

Depending on what shape of home range is assumed, the home-range prediction allows a crude approximation of lateral distance normally traveled. These estimates are of use in assessing the

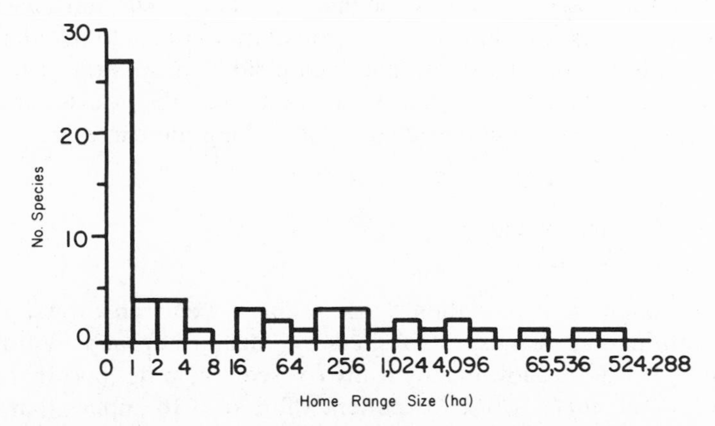

Figure 8.7 Frequency distribution of estimated home-range sizes for cursorial mammals of western Oregon.

magnitude of the interisland distances relative to the species of the region. For example, assuming an elliptical home-range shape (major axis = 2 × minor axis), thirty-six species have a normal traveling distance of 1,200 feet (360 m) or less. There is no other fortuitous breaking point in the frequency distribution of home-range sizes that might be useful in specifying either island size or interisland distance (fig. 8.7).

Home-range size projections can also be used to establish the theoretical maximum for species-area curves for western Oregon mammals. Species are first ordered in terms of increasing home-range size. From this, it is easy to predict that no species can be contained in a patch smaller than the size of the smallest home-range projection. The area associated with the fifth species would be the minimum necessary for five species; the area corresponding to the tenth species would be not only the minimum necessary for that species, but the minimum necessary to contain totally at least one individual of ten species. When the cumulative number of species is plotted against associated areas, the maximum potential species-area curve is obtained (fig. 8.8). Because this projection is based on single individuals and because there is no inference about habitat features, it is the outside limit for mammal species-area curves. The area below the curve represents the feasible domain for all real-world curves; all species-area curves derived from empirical research should fall below it. For example, if we agree that two individuals of a species have a combined home range greater than or equal to one individual, then the home-range projection for that species will be above that given in table 6.3. The grizzly and a pack of wolves have been added to the projection in order to maintain the integrity of the species list and the outside limits on the curve.

Two important lessons can be learned from this species-area relation. The least-squares equation describing the curve is:

$$S = 16.3A^{0.16}$$

where S = no. of species
A = area in ha

The exponent of this equation is remarkably close to the lower limit for continental islands established by MacArthur and Wilson (1967). It can be shown analytically (or seen from fig. 6.4) that a species-area curve with an exponent value of 0.16 implies that a seventy-six-fold increase in area would be necessary to double the

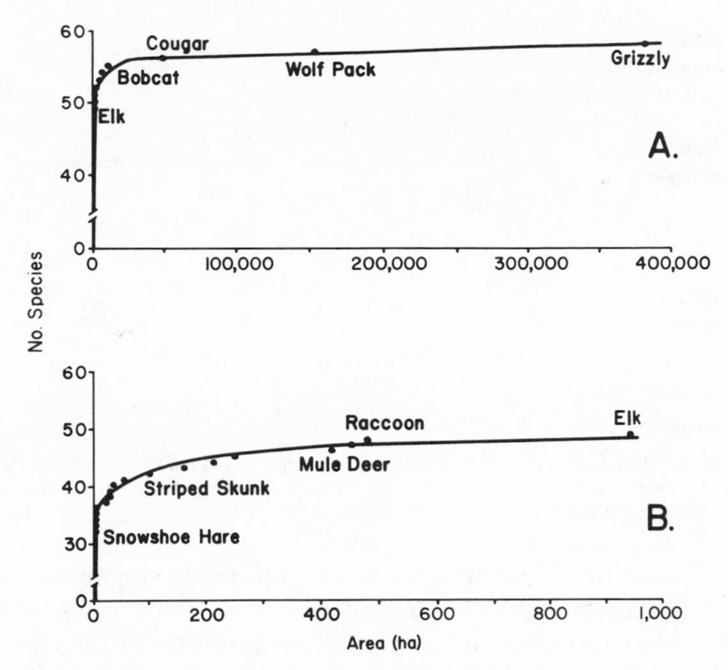

Figure 8.8 Theoretical maximum species-area curve for western Oregon mammals based on the cumulative number of species whose home-range size is encompassed by any given area. B is simply a higher resolution extension of A. The area below the curve represents the feasible domain for empirical species-area curves.

number of species. Thus, if some patch of ten acres supported thirty species of mammals, a patch of 760 acres would be necessary to support sixty species.

Another way of using the same information involves the number of species potentially added for each tenfold increase in area. Whereas very small additions in area may increment large numbers of species when the patch sizes are small, very large increments in area are necessary to add a small number of species when the patch size becomes large (fig. 8.9).

Since none of the mammals considered here nor any known vertebrate is totally restricted to old growth, the species-area relation does not apply just to old-growth habitat patches. Nonetheless, the material is highly relevant to those who would argue in favor of setting aside large areas in the belief that an entire faunal community could be contained therein. Even if the full complement of large

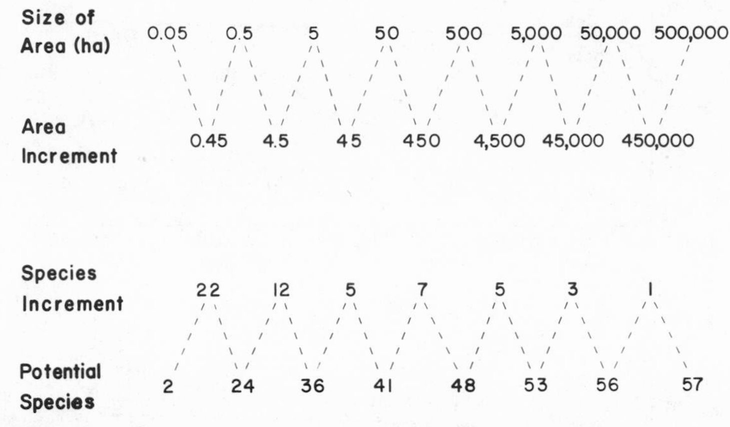

Figure 8.9 Relation between the number of western Oregon mammal species added and the cumulative number that could theoretically be supported by increasing the size of a habitat patch.

carnivores still existed in western Oregon, the area necessary to contain them would be on the order of 1,500 square miles (4000 km^2). The above calculation pays no regard to the issue of minimum viable population size. The occurrence of large mammal species in national parks of the western United States supports the notion that species only occur in parks that are much larger than the calculated home-range sizes given here. And herein lies the major distinction between the purpose and strategy of a park and wilderness area system as opposed to managed production forests. Parks, preserves, and wilderness areas must be considered in light of how well they do or do not function as natural ecosystems and preserve entire faunal assemblages. In this sense it is appropriate that the faunal community be considered a subset of the natural system components within the preserve. On the other hand, habitat islands in a managed forest cannot realistically be expected to contain totally an entire faunal assemblage. In this situation, each old-growth island must be considered as lying within the home range of the wider-ranging species. In this sense, it is the habitat islands that are the subset within the species ranges. I believe this distinction to be of critical importance in evaluating conservation strategies, and it will be discussed at greater length in the final part of this book.

PART 4

A Planning Strategy

9

A System of Long-rotation Islands

The previous chapters have detailed the relevant states of nature, described the most relevant animal community ecology, demonstrated the applicability of island biogeography principles, and discussed the criteria for assessing and evaluating the objective of maintaining biotic diversity. I shall now develop a planning and management strategy that can be applied to the specific tract under analysis or extrapolated to other areas. It is but one strategy; however, I believe it to be the best one. It builds upon two premises: (1) that biotic diversity will be maintained on public forest lands only if conservation planning is integrated with development planning; and (2) site-specific protection areas must be designed so they function as an integrated landscape system. These are consistent with recommendations of the U.S. Strategy Conference on Biological Diversity (U.S. Dept. State 1982). At minimum, this approach shifts much of the emphasis away from the old-growth system and puts it on the system of long-rotation and/or old-growth islands.

Any functioning, opening system can be shown to consist of components, with attendant linkages, inputs, and outputs. If the paleoecological record and analyses of recent ecological changes on montane islands and in isolated reserves are correct, then we can anticipate future inputs of biotic diversity to the old-growth island system to be minimal. It is hoped that the outputs will be positive. The individual components will have analyzable characteristics in their own right, but analysis of the individual components is not essential either for describing or understanding the larger system. In discussing the issue of old-growth management, Overton and Hunt (1974, 334) appropriately noted that "we can't properly make a decision whether or not to cut a particular piece of old-growth unless we know a great deal about the relationship of this particular piece to the rest of the forest." Thus consideration of the old-growth island system should consist of (1) definition and description of the components, i.e., old-growth habitat islands; (2) the size-frequency distribution of these islands; (3) the spatial distribution pattern; (4) guidelines for the choice of one old-growth island over another (i.e., location and expected contribution); and (5) description of the connectivity or linkages between components. These are discussed in some detail below, but I first introduce the concept of long-rotation forest islands.

Long-rotation Islands vs. Old-growth Islands

Because of our lack of knowledge about intricate old-growth ecosystem relations (see Franklin et al. 1981), and the notion that oceanic islands never achieve the same level of richness as continental shelf islands, a major commitment must be made to set aside representative old-growth ecosystems. This is further justified because of the lack of sufficient acreage in the 100- to 200-year age class to serve as replacement islands in the immediate future. These old-growth islands must not only serve faunal conservation interests at present, but must ensure a representative and viable gene pool and maintain the complex, functioning ecosystem in perpetuity. If these old-growth ecosystems are expected to function independently of their surroundings, then their size might need to be so large that their conservation would be outside the realm of existing conditions or current political and economic restraints. As discussed earlier, a way to moderate both the demands for and the stresses placed upon the old-growth ecosystem, and to enhance each island's effective area is to surround each with a long-rotation management area.

There is a second and perhaps more important reason for considering long-rotation islands as the basic design component. Although we should pursue the objective that old-growth ecosystems will be protected and maintained in perpetuity, perpetuity is a long time. Natural destructive forces will be at work and events that are highly improbable in the short run may be inevitable in the long run. Since our reason for maintaining old growth is not so much aimed at the present as it is at the future, we must follow procedures that will ensure the sustained development of replacement stands. We can ensure that these replacement stands have the maximum probability of colonization by placing them next to existing old growth. This gives the added advantage that long-term planning can assume the location of long-rotation islands to be more of a constant than a variable.

The combined advantages of this system of long-rotation management islands are as follows:

1. The value of a patch of habitat is not only dependent upon its structural characteristics (its content), but also upon the matrix within which it occurs (its context). The degree of similarity between the vertebrate community of old growth and mature timber is very high (97%). The degree of similarity between old growth and early to mid-successional stages is low (42% and 46%, respectively). From a habitat standpoint, a patch of old growth that is surrounded by mature timber is less distinct than a patch surrounded by regeneration areas. The surrounding mature but managed stands increase the effective size of the old-growth island.
2. The strip of surrounding forest that is necessary to buffer changes in wind, temperature, light, and relative humidity within the stand can, and perhaps should, be managed forest. Ensuring that the majority of this buffer is mature will effectively reduce the minimum size of old-growth stands necessary to maintain ecosystem processes and structural integrity (e.g., prevent the rapid drying out of down logs and snags).
3. Whereas regeneration stands are most susceptible to fire, mature stands are least susceptible. Surrounding old-growth islands with mature stands will provide the best buffer against lightning fire starts and the intrusion of fire and destruction of the old-growth island.
4. Since surrounding long-rotation management stands will increase the effective size of individual old-growth patches, more emphasis can be shifted to increasing the number of islands.
5. The greater number of islands that will be possible will facilitate conservation of endemic species in unique habitats, and the clinal genetic variation and/or diverse ecotypes within species.
6. Because of increased effective size of individual islands and the resulting increase in number of islands, the average interisland distance and average nearest-neighbor distance will be greatly reduced.
7. The increased number of islands, geometrically increased two-way interactions between islands, and geometrically reduced interisland distances will facilitate linking the islands together with corridors. This will greatly increase the prospects for immigration and genetic interchange.

8. Increased number of islands, increased linkage or connectivity, and increased emphasis on the interconnected system of islands as opposed to the old-growth ecosystem itself will be more effective in conserving large carnivores and wide-ranging species. Conservation of the top carnivores is no doubt worthwhile in its own right, but encouraging their presence and movement throughout the system will facilitate the positive effects they may have in balancing ecosystem function.

9. A larger number of islands, with smaller interisland distances and greater connectivity, will greatly increase the prospects for frequent colonization by species that may have disappeared from any given island.

10. The greater connectivity between islands will facilitate the prospects of occasional genetic interchange between isolated demes of a population. Since each individual carries 50% of the parent population's variability, occasional immigration will help keep the subpopulations viable.

11. A system consisting of more islands may allow higher numbers of territorial old-growth species to occur, since each island may serve as the defended territory nucleus. (This does not imply that the entire territory need be contained within the island.)

12. Containing each old-growth ecosystem within a long-rotation island will greatly simplify the planning associated with maintaining a spatial pattern of islands since replacement stands will always occur in close proximity to where the old growth is presently located. It will also ensure that a known combination of replacement stand ages will occur adjacent to the original old-growth island.

Long-rotation Island Characteristics

Current evidence suggests that, depending on the site, from 175 to 250 years have been required for old-growth forests to develop in the recent past (Franklin et al. 1981). I have assumed that 240 years, which is three times eighty years or approximately three short-rotation periods, will also suffice in the future. From this, a long-rotation island is defined as a management unit consisting of a current old-growth stand and several surrounding adjacent stands managed on a 320-year rotation. This schedule implies that three short-rotation periods are required to achieve old-growth status and one additional short-rotation period will allow the stand to function as an old-growth ecosystem. It implies that under equilibrium conditions 25% of long-rotation island acreage would consist of old growth and 75% would consist of recruitment stands. From a wildlife conservation standpoint, the entry interval between forest operations should be long yet should allow necessary silvicultural options. If the island consists of an old-growth core area and nine replacement stands, then the required entry interval would be about thirty-five years. Assuming a clearcut and planting silvicultural system with two thinnings per short rotation, entries for clearcutting one stand and/or thinning one or more stands would occur every seventeen years (fig. 9.1).

Edge-effect principles such as discussed by Thomas and his coauthors (1978, 1979b), and Harris and McElveen (1981) should be considered when deciding the sequence of stands to be clearcut and thus the overall distribution of stand ages. It seems obvious that neither a mature stand occurring next to old growth nor a clearcut next to a clearcut creates a substantial edge effect for most species (fig. 9.2). A clearcut juxtaposed with a dense, mature stand creates maximum edge distinction in the immediate time frame, but if the mature stand were cut at the next entry period, there would again be little edge distinction between the two regeneration stands. It

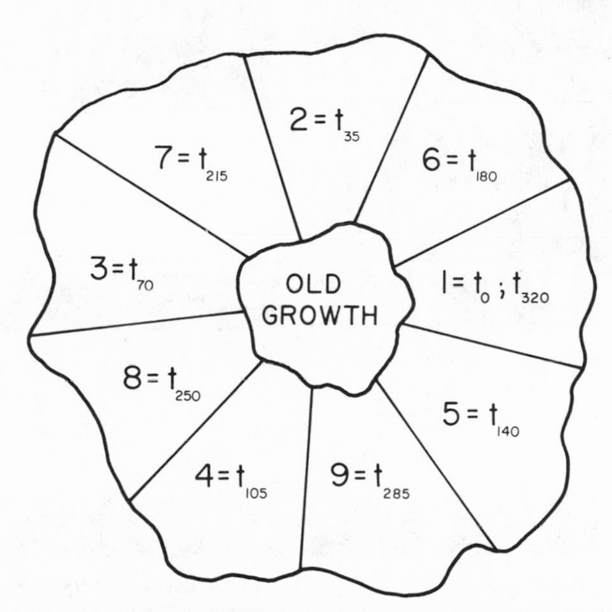

Figure 9.1 Schematic illustration of a long-rotation island illustrating the cutting sequence (alternate stands) that leads to maximum average age difference between adjacent stands over a complete cutting cycle.

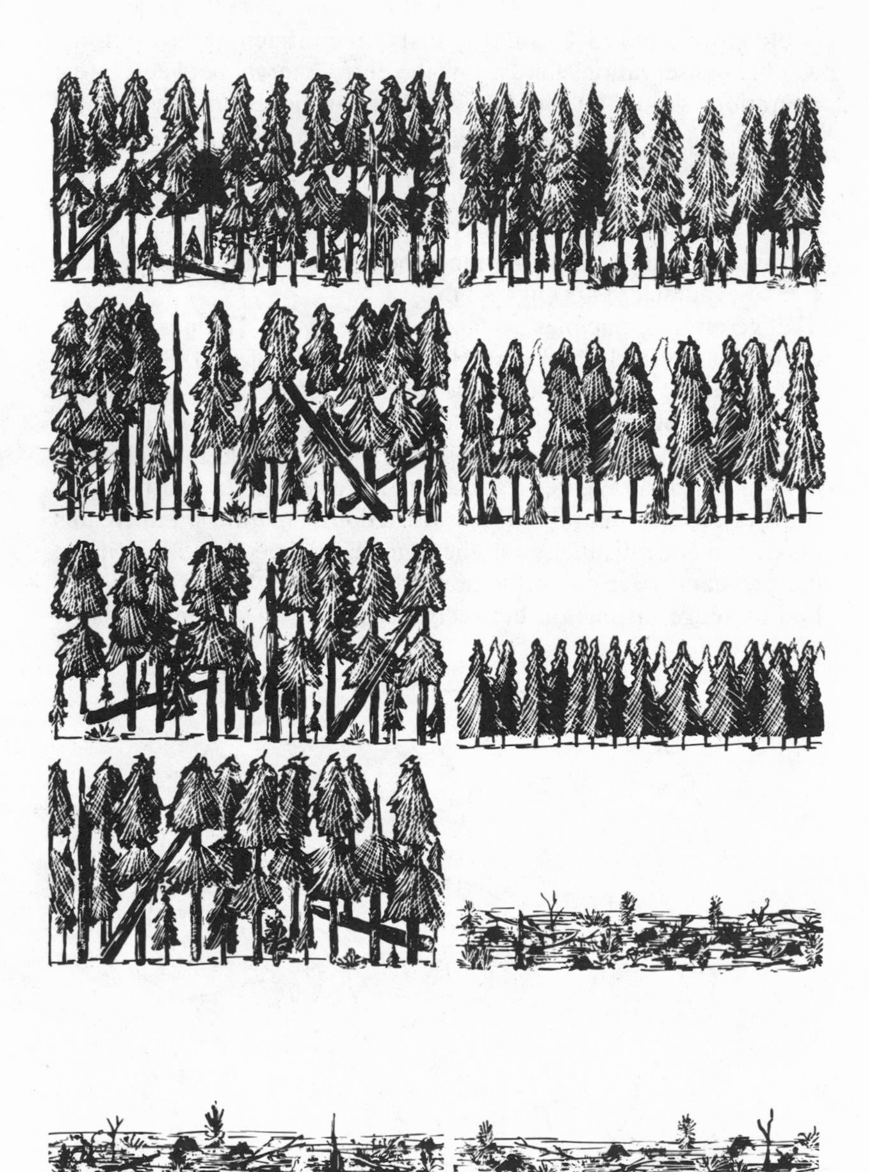

Figure 9.2 Edge effects are believed to be directly related to age differences between adjacent stands. Little edge effect occurs between old growth (upper left) and large sawtimber (upper right) or between two recent clearcuts. When summed over a complete rotation, maximum edge effects will occur by putting clearcuts next to mid-rotation-aged stands.

can be shown algebraically that the maximum amount of age difference between stands that will occur over the course of one or more cutting cycles will be achieved if the clearcuts are always positioned next to a mid-rotation-aged stand (fig. 9.3). Therefore, since the long-rotation is defined as 320 years, maximum edge effect will occur when clearcuts are positioned next to stands approximately 160 years of age. Although this is not the place for discussion, it is important to recognize that all edge effects do not have positive benefits for native fauna. For this reason, the argument that old-growth islands should be totally surrounded by regeneration areas in order to maximize edge effects should be substantially discounted (e.g., see Robbins 1979; Brittingham and Temple 1983).

Cutting patterns and rules for scheduling can be derived from the information presented above. Under equilibrium conditions, long-rotation islands consisting of an odd number of stands always have a greater average age difference between stands than do islands consisting of an even number of stands (fig. 9.4). The average age difference between stands increases with increasing stand number but approaches 50% of rotation age as an asymptote when the

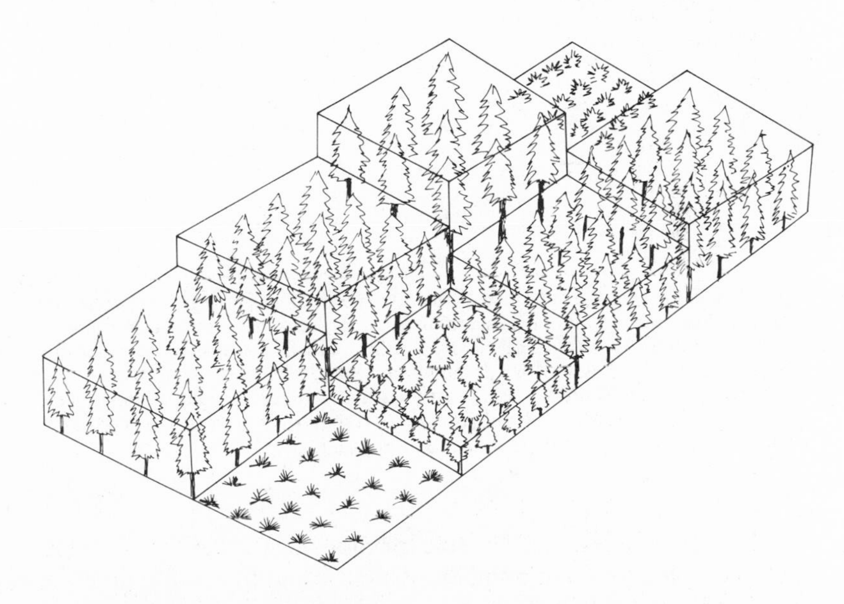

Figure 9.3 When integrated over two or more rotations, the maximum age difference between adjacent stands will occur if clearcuts are placed next to stands of mid-rotation age.

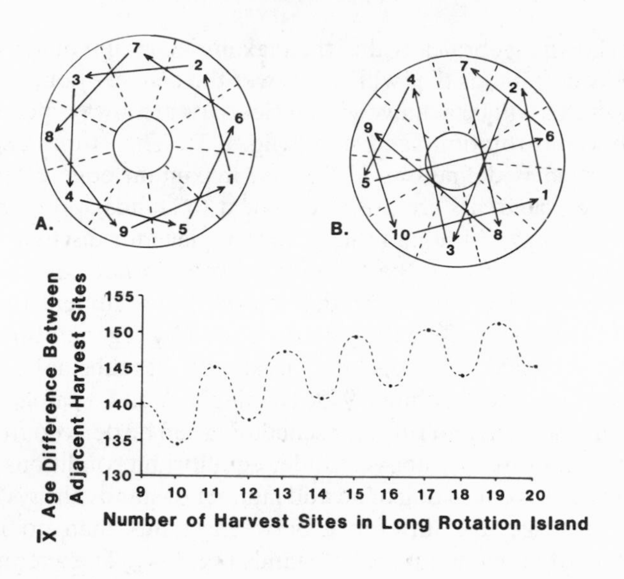

No. Harvest Sites	$\bar{X}$ Age Between Adjacent Sites	No. Years Between Successive Cuts
9	140.0	35.0
10	134.4	32.0
11	145.0	29.0
12	137.8	26.6
13	147.6	24.6
14	140.6	22.9
15	149.3	21.3
16	142.6	20.0
17	150.6	18.8
18	144.2	17.8
19	151.6	16.8
20	145.6	16.0

Figure 9.4 A long-rotation island with an odd number of stands (e.g., nine, as in A) simply requires cutting alternative stands in a counterclockwise direction. Assuming a long rotation of 320 years, the mean age difference between adjacent stands is always greater for islands with odd-numbered stands than for those with even-numbered stands and increases as a function of the number of stands in the island. The cutting sequence for even-numbered stands is somewhat more complex (B), but involves cutting the second stand to the right, return to previous cut, then second to the left, etc.

number of stands is large. The rule for choosing the next site to cut is more complex for even-numbered stands than for odd-numbered stands (fig. 9.4). Based on the above information, it is recommended that a long-rotation island consist of an old-growth core area surrounded by nine replacement stands. After equilibrium is reached, the initial old-growth core area can be cut with no loss to the island system. Whether or not the old-growth core area is ever cut, the long-rotation island described will ensure that 66% of the surrounding buffer stands are over 100 years old, while the 33% regeneration area will provide forage and habitat for early successional species (fig. 9.5). Based on what is known about tropical communities, this consideration of early seral stages may be critical (L. E. Gilbert 1980).

Island Size Frequency Distribution

All previous discussions of island sizes have either implied average island size or have dealt with an empirical size distribution such as that of the Siuslaw National Forest. It is important to bear in mind that there are three characteristics that determine size frequency distributions: the type of distribution, a measure of central tendency (e.g., the mean), and a measure of dispersion (e.g., the variance). I do not know of biogeographic arguments to compel any particular distribution when designing a system of habitat islands. On the other hand, the log-normal distribution is so widely applicable throughout ecology as to make it appealing (Hutchinson 1953; Preston 1962a; May 1975). The size-frequency distribution of animals is graded so that an inverse relation between size and abundance usually exists within ecological communities. Daily energy demand is a function of the body size of individuals, and therefore home-range size is a power function of body size. Finally, the energetic relation of different trophic levels coupled with home-range size allows progressively fewer organisms to occur at succeedingly higher trophic levels. These three basic relations—abundance, trophic position, and spatial movement—suggest that some geometric (e.g., log-normal) distribution may be especially applicable to the animal community. A second line of reasoning for choosing a log-normal island size frequency distribution derives from the landscape itself. Horton (1945), Strahler (1957), and others have shown that watershed areas and the size-frequency distribution of stream lengths of different order follow geometric relations. The size-frequency distribution of fires (U.S.D.A. 1980) and perhaps other natural events such as windstorms are also geometric. Thus there are natural mechanisms to recommend a graded relation between size and frequency of old-growth habitat islands. Because of these and other intuitive reasons, I have chosen

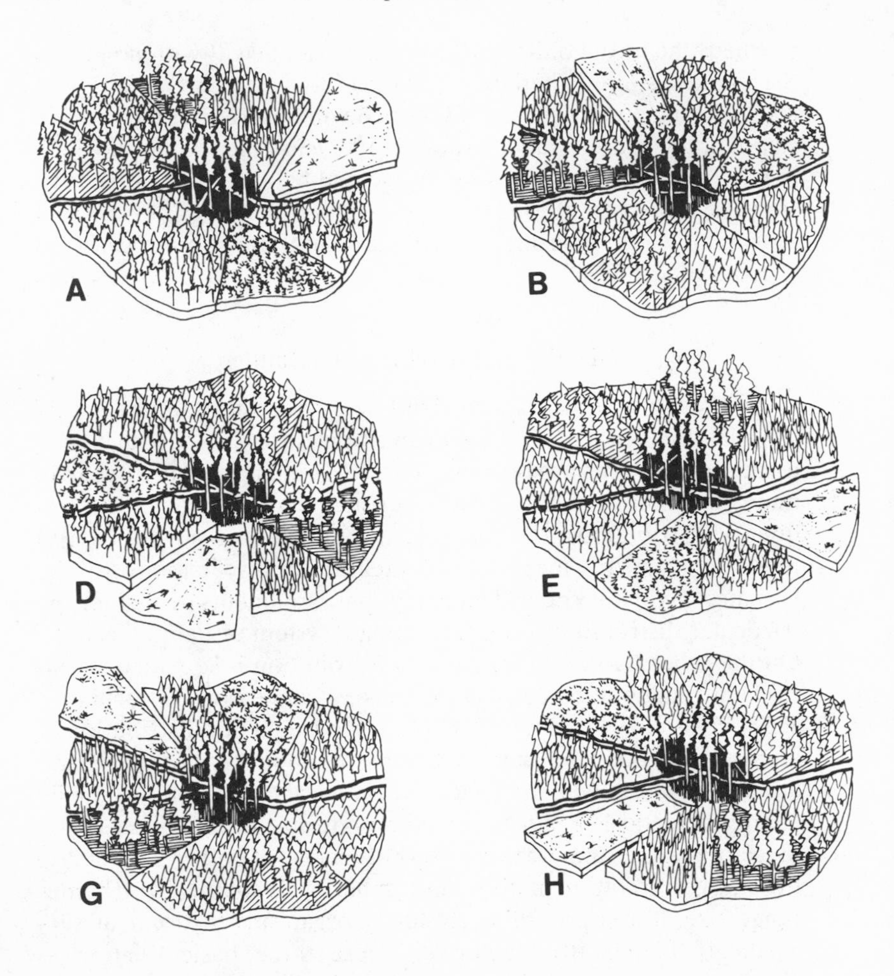

to use a log-normal frequency distribution for old-growth island sizes. Examples of calculations for the Willamette National Forest are given (in figures 9.6 and 9.7). The Willamette National Forest contains approximately 1,300,000 acres (5,263 km^2) of commercial forest acreage. As described above, a commitment of 20% of commercial forest acreage (260,000 acres, 1,052 km^2) to long-rotation islands will insure 5% (65,000 acres, 263 km^2) in the 240- to 320-year age class in perpetuity. As an example, if a log-normal frequency distribution were accepted, 100 old-growth islands with an average size of 650 acres (263 ha) could yield the distribution shown in figure 9.6. There would be 12 islands of about 50 acres (20

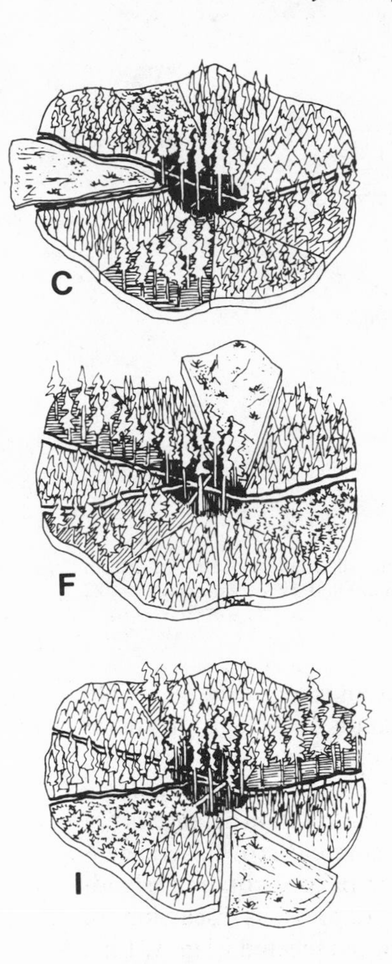

Figure 9.5 An old-growth patch surrounded by a long-rotation island that is cut in a programmed sequence such as that shown in figure 9.1.

ha) each, 20 islands of nearly 200 acres each (80 ha), 25 islands of about 760 acres (310 ha), and a decreasing number of large islands. It is important to note that there could be one island in each of the very large size classes ranging to about 31,000 acres (125 km^2) each.

If the same acreage commitment (260,000 acres to long-rotation management) were accepted but the mean and variance were changed, the distribution would change accordingly (fig. 9.7). For example, 800 islands with an average acreage of 82 acres (33 ha) could still produce over 500 islands with average size of 160 acres (65 ha) each and 300 islands in the large size classes.

The above calculations assume a commitment of 20% of com-

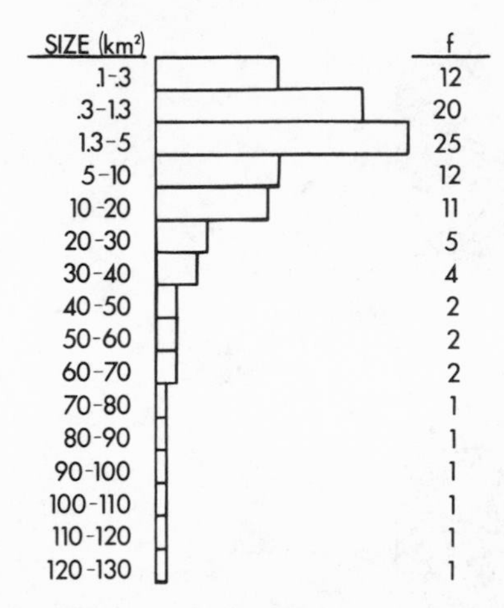

Figure 9.6 An example of log-normal size-frequency distribution of 100 old-growth patches assuming an average size of 650 acres (263 ha) and a wide variance, $a = 0.2$ in the formula: $f_r = f_o e^{-(ar)^2}$.

mercial forest acreage to long-rotation islands, ensuring 5% in old growth at any given time. Assuming clearcuts to average forty acres (16 ha) and nine replacement stands in each long-rotation island, approximately 360 acres (146 ha) of long-rotation stands would be associated with each island. If the total long-rotation acreage is not to exceed the 20% commitment and the core old-growth acreage is not to fall below 5%, then the number of long-rotation islands should not exceed 550. In conjunction with previous discussions of the advantages of a large number of interconnected islands, I believe this to be a reasonable number. Therefore, given that (1) the long-rotation island concept is chosen so that old-growth stands are buffered by mature timber; and (2) 20% of commercial forest acreage is committed to long-rotation management; and (3) a frequency distribution that includes a significant number of large old-growth stands (several hundred acres) is applied, I conclude that approximately 500 long-rotation islands containing a core patch of old growth with an average size of 130 acres (53 ha) is the superior alternative for the Willamette National Forest. Since this calculation assumes that most of the perimeter of the old-growth stands (80%; see fig. 8.2) will be buffered by mature timber,

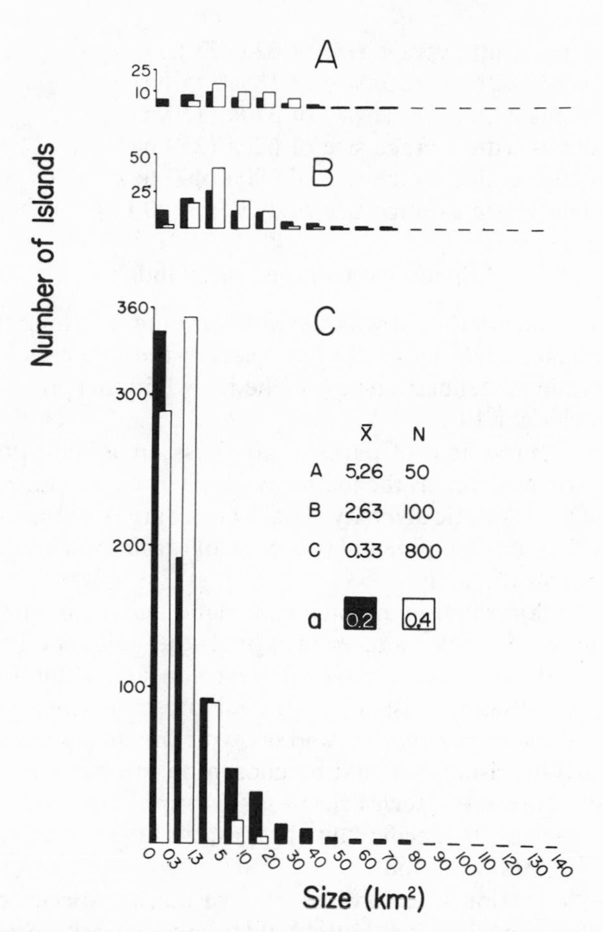

Figure 9.7 Three island-size frequency distributions (panels A, B, and C) based on different means and variances. Solid bars represent the distribution with the *larger* variance ($a = 0.2$) and open bars represent the *smaller* variance ($a = 0.4$). In all cases, A, B, and C, the product of island number and average size totals 263 square kilometers (102 mi^2).

acreage being committed to long-rotation management should be designated from existing old-growth forest. That is, 260,000 acres of old growth should be committed to long-rotation island management with scheduled cutting reducing the old-growth acreage to 65,000 acres by the year 2255. An example of 495 log-normal old-growth islands with an average size of 122 acres (50 ha) which total only 60,625 acres (245 km^2) is given below.

400 islands with average size of 62a (25 ha)
50 islands with average size of 185a (75 ha)
25 islands with average size of 370a (150 ha)
10 islands with average size of 620a (250 ha)
5 islands with average size of 990a (400 ha)
5 islands with average size of 1,235a (500 ha)

Spatial Distribution of Islands

Although the details of which stands to commit to long-rotation management should be left to biologists at the forest and district level, certain guidelines can be specified. (Additional principles are given in chapter 10).

The present system of parks, wilderness areas, and preserves should be considered as the mainstay of any program aimed at the maintenance of biotic diversity. These larger areas will serve as the closest thing to a "continental source pool" from which immigrant species might originate.

Contribution to an interconnected island system should be the first criterion for selection, even though the uniqueness of individual stands must be considered (Overton and Hunt 1974). A functioning system of islands will best preserve overall species richness, endangered species, and ecotypic or clinal genetic diversity. Individual islands should be chosen on the basis of their (1) location; (2) overall species richness; (3) support of endemic and disjunct species or specific faunal types; and (4) probable occurrence of genetic variation.

Lower elevation sites have greater vertebrate species richness than higher elevation sites, and should be given priority. Site potential indices (e.g., site index) probably have as much relevance to wildlife productivity as they do to timber productivity. Old growth on low potential sites is not an acceptable substitute for high site-potential areas.

Security and protection are important considerations. The system of old-growth ecosystems should not be designed on the basis of its contribution to the present, but on its contribution to the future. Old-growth ecosystems that can be protected long into the future should be favored over those that cannot.

Design patterns fitted to the landscape are superior to those that look appealing in text books and brochures. This topic is discussed below.

Travel Corridors and Connectivity of Islands

As mentioned earlier, an important distinction between habitat islands and true islands is that true islands draw species from a "mainland" system while the habitat island system will probably not have a secure source pool to serve as a "continent." This means that travel corridors and interconnections between islands take on added importance for habitat island systems. Travel corridors that connect forest fragments are believed to enhance species richness of breeding birds (MacClintock et al. 1977) and they represent an established wildlife management technique in some regions (Gehrken 1975; Buckner and Landers 1980). In midwestern states, fencerows and other "environmental corridors" increase dispersal of climax tree species and affect tree species distributions by influencing the travel of seed-dispersing wildlife species (Boucher 1979; Sork 1979; Levenson 1981). The role of island interconnections is sufficiently well establihed to lead some authors to conclude that "derivation of diversity-maximizing formulations for biological populations across time and across a region . . . in essence . . . involve the maximization of the interaction between islands . . ." (Rudis and Ek 1981, 253). Indices for quantifying the degree of connectivity between components are available (e.g., Pielou 1979).

One approach to enhancing the connectivity of the system might strategically locate the smaller old-growth patches so they serve as "stepping-stone" islands between larger centers of dispersal. Although this approach should be drawn upon, it is inferior to physically continuous corridors. The value of smaller stepping-stone islands will be greatly enhanced by choosing them with a pattern in mind and using corridors to couple the entire system.

Continuous extensions of closed-canopy forest of sufficient width to resist blow-down would be ideal. In many instances, sufficient expanses of old growth remain so that a residual system of islands and interconnections could be designed. Alternative interconnections of perhaps less but still considerable merit would involve scenic roadside strips, ridge systems, and strips of mature sawtimber. An alternative I believe superior to the above is the use of riparian strips. These ecosystems are sufficiently noteworthy to be discussed in greater detail.

The riparian strips of the western Cascades usually occur as distinct ecosystems containing permanent water and are dominated by deciduous hardwood species. Over and above, but not independent of, their function as anadromous fish habitat, they rank near the most productive of all habitat types in the western Cascades. Riparian strips are utilized by more species of wildlife than any of the twenty-nine other habitat types in California (Anon. 1966, 53; Gaines, 1980). Voth (1963, 97) observed that "this represents the location of the most numerous fauna. . . ."

There are many reasons why riparian communities represent such ideal wildlife habitat. Some of these are general and apply to many species while others are specific and apply to individual species. Perhaps the most fundamental reason derives from the fact that riparian ecosystems receive water, nutrients, and energy from surrounding "upstream" systems. These additions not only allow greater richness, but also impart more constancy to the occurrence of resources such as water. In the Cascades, permanent surface water occurs only in lakes and in streams. It is this water that allows numerous amphibian and aquatic bird and mammal species to occur. In addition to this direct effect, the presence of water makes possible the presence of fish and other aquatic organisms that form the bases of several food chains. Thus Ingles (1965, 81) states that the insectivorous mammals of the "Pacific Coast are nearly all inhabitants of damp, moist places." Entire sequences of terrestrial vertebrate species occupy similar functional niches along different-sized streams in accordance with the river continuum concept (Vannote et al. 1980). For example, a series of carnivorous, amphibious mammals that play similar ecological functions utilize different food particle sizes, different stream orders, and occur at somewhat different elevations along the stream gradient. These include the northern water shrew, marsh shrew, mink, otter, and grizzly (fig. 9.8).

The second proximal reason that riparian strips provide superior wildlife habitat derives from hardwood dominance. By producing nectar, catkins, fruits, and seeds coevolved with animals for dispersal, angiosperms provide a food base unavailable from conifers. Species such as alder provide the most preferred forage for the white-footed vole and thus the habitat of this vole is largely limited to riparian strips. A study of ruffed grouse habitat in Wisconsin found that "the aspen-alder habitats supported a mean annual density of 9.8 drummers. Aspen stands of all ages devoid of alder, however, supported a mean density of only 2.0 drummers/40 ha" (Kubisiak et al. 1980).

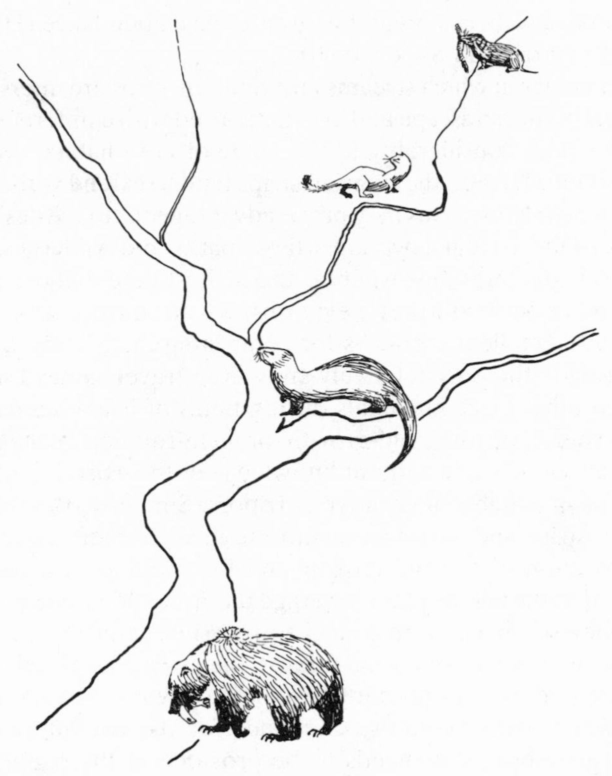

Figure 9.8 Association of different-sized carnivorous mammal species with stream order and typical food particle size in accordance with the stream-continuum concept.

Hardwood and riparian habitats seem to support higher populations of birds throughout the year. These additional birds consist of insectivorous summer residents as well as numerous granivorous and omnivorous permanent resident species that remain throughout the year. This means that the ratio of summer resident to permanent resident species is much higher in hardwood than in coniferous habitats (Dirks-Edmunds 1947; Bratz 1950; Bowles 1963; Anderson 1970). It appears that a higher insect abundance in the hardwoods complements the greater fruit and seed production to allow these higher bird densities (Blake 1926; Dirks-Edmunds 1947; Hamilton 1962). A greater number of cavities and the different branching structure may lead to higher foliage height diversity which may also help explain the greater bird abundance (Harris et al. 1979; Harris and Skoog 1980).

The manner in which streams and riparian strips are interspersed throughout the landscape and are juxtaposed with coniferous forest probably adds considerably to the value of both habitat types. In mountainous terrain, the riparian strips tend to extend vertically up and down the slope. This may prove advantageous to the design of a habitat island system since most large parks and wilderness areas occur at high elevations, whereas the highest potential old-growth sites tend to occur at lower elevations. Riparian strips would seem to provide excellent corridors for connecting the islands. Streams may provide the only relatively snow-free travel routes for deer, elk, and other large mammals during times of heavy snowfall.

The choice of ideal old-growth or long-rotation management areas cannot be made without knowledge of the existence of riparian strips or suitable alternative corridors. Similarly, the choice of riparian strips and corridors cannot be made without an approximate location of the old-growth and long-rotation islands. This implies that during the planning stage there should be close interactive review of what pattern is ideal and possible at the regional level, with what is ideal and available at the forest level. The same interactive review must occur between the forest supervisor's office and district planning staffs. Guidance for the overall system of islands and connectors needs to be provided at the regional and then forest level, but decisions regarding specific stands and corridors need to be made at the district level. Without integration into present planning operations and the landscape, the proposal will not work and the system will not persist. Yet, without higher-level guidance, there will be no overall system.

10

Fitting the System to the Landscape

At least four major questions should be asked when deciding which stands to allocate to long-rotation management: (1) Is the stand strategically located with regard to the function and integrity of the overall system? (2) Does the stand make a specific contribution to genetic diversity in terms of ecotypes, endemic species, or greater species richness? (3) Is the choice strategic in terms of present cutting trends and plans (i.e., does rejection foreclose critical future options? (4) Do the stand and its linkages fit into the landscape pattern and process? For example, is it a high-risk or relatively secure location? Information regarding the first two points has been presented. The third point can only be addressed by local or regional authorities. The issue of landscape form and function will now be considered briefly.

Although adapted to local site conditions, the distribution of plants and animals reflects a highly dynamic veneer imposed on a rugged physiographic and topographic background. The fact that Douglas fir, which is generally thought to be a shade-intolerant seral species, can dominate the landscape indicates how dynamic the destructive forces have been. By analogy, ecosystems are like flesh on a skeleton with their stability and function largely determined by location. Lower-elevation sites tend to be richer because of time stability and perpetual nutrient and energy subsidies from systems higher in the watershed. Ecosystems on some sites are clearly more prone to destruction by fire, landslide, insects, and disease than those in other landscape positions. The presence or absence of a species may be more dependent on topographic position or aspect than on the age or condition of the forest occurring there. Researchers such as Swanson (1980, 1982) and Miller (1978) are beginning to relate ecosystem analysis and resource management to geomorphology and the landscape.

Descriptive correlations between form and function are well established in the basic sciences such as botany and zoology (Thompson 1961; Portman 1969). For example, given relatively similar ecological niches in geographically isolated ecosystems (e.g., Australia and North America), organisms of greatly different evolutionary background frequently develop remarkably similar characteristics. Although no usable synthesis or application above the organism level has been made, the approach of relating form to function at the landscape level seems clear and rewarding. Given that we can clearly articulate the ecological function required of the habitat island system, perhaps we can draw upon basic design forms to achieve it (Harris and Kangas 1979).

The dendritic pattern is a form that is common at various hierarchical levels in the plant and animal kingdoms. It is common at the cellular level of the dendrite-axon, the organ-system level of circulatory systems, and the organism level of plant root and branch systems. It is commonly recognized in drainage and river systems. The function associated with this form is usually the collection, concentration, or distribution (e.g., tributary vs. distributary systems) of energy, nutrients, or other materials (water or gas). The dendritic pattern is obvious throughout the landscape of the western Cascades but a logical extension of the basic form-function relationship at the landscape level remains to be made.

First-, second-, third-, and higher-order stream systems (Horton 1945; Strahler 1957) of the Cascades have been mapped and appear highly dendritic (fig. 10.1). The concentration of water from a dispersed state to progressively more concentrated central rivers results. Over millennia, erosion has cut the valley systems into the same pattern, and therefore it has been convenient to construct road systems that conform to that same basic pattern (fig. 10.1). To suggest that anadromous fish are limited to the river system seems trite, but to accept that this important renewable resource is distributed in accordance with the same underlying dendritic pattern is not trite. It follows that much of the potential of fisheries and water-based recreational resources adheres to the same pattern. In the western Cascades, where little precipitation occurs during the summer, the distribution of a large number of terrestrial vertebrates is also closely related to the dendritic stream pattern (e.g., fig. 9.8). Because fire incidence is correlated with proximity to roads and amount of human activity, the frequency and threat of fire are related to the road system. This has direct bearing on the

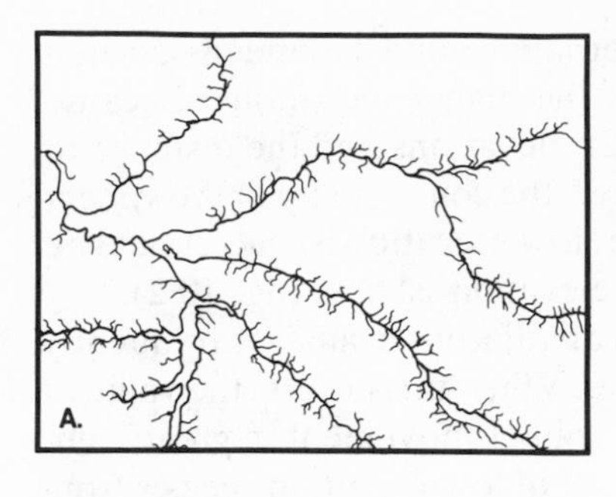

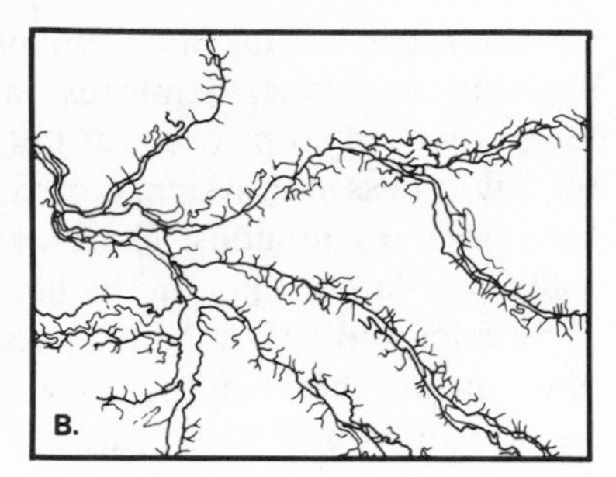

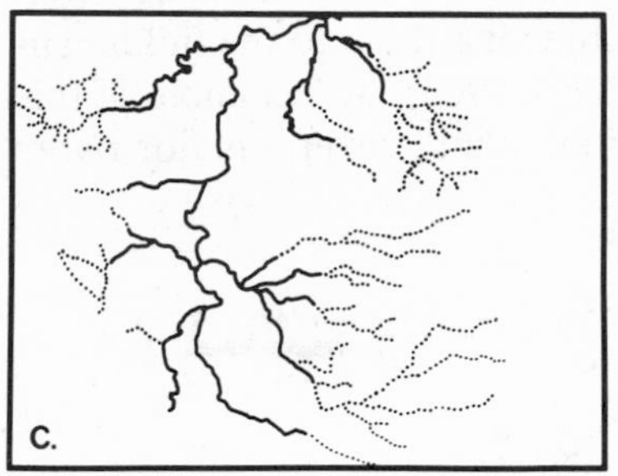

Figure 10.1 Dendritic pattern of (A) streams and rivers and (B) road system in the Oak Ridge district of the Willamette National Forest. Anadromous fish runs (——) and spawning areas (. . . .) (C) illustrate that the distribution pattern of many biological resources is also dendritic.

ideal location of long-rotation management areas. Thus for protection and for other reasons they should be located away from major human traffic arteries or nodes.

Riparian strips follow the stream and river channels; therefore the smaller "dendrites" occur at the higher elevations whereas the larger riparian forests are associated with the lower reaches. If the riparian strip system is to serve a major role in interconnecting the long-rotation islands, then leave strip widths should be scaled to stream order. The narrower leave strips should be associated with the smaller, low-order streams at higher elevations while wider leave strips should occur along the rivers at lower elevations. This not only makes sense from the standpoint of stream management but also for biogeographical reasons. The riparian strips are expected to serve as wildlife travel corridors, and since the larger preserves occur at high elevations, the funneling of individuals and species from higher to lower sites may be important. Wider strips of riparian vegetation along the larger streams and rivers will help compensate for increased human use and the fact that the low elevation sites are farther removed from the large reserve system. If long-rotation islands are placed near riparian strips there will be opportunity for many more (but perhaps smaller) islands associated with the abundant "dendrites" at the higher elevations. Because fewer riparian strips occur at lower elevations and the distance to the wilderness area system is greater, the connectivity of the system becomes more tenuous. Therefore, low-elevation islands should be somewhat larger than the higher-elevation islands (fig. 10.2).

Streams and occasionally rivers of adjacent drainages frequently abut one another near the ridge top. When this occurs, the value of the stream course as a route for animal movement is greatly enhanced as it facilitates the movement of low-elevation species from one drainage to another without subjecting them to the full harshness of high-elevation conditions (Voth 1963). An example of this occurs in the Mount Jefferson region, where several major rivers

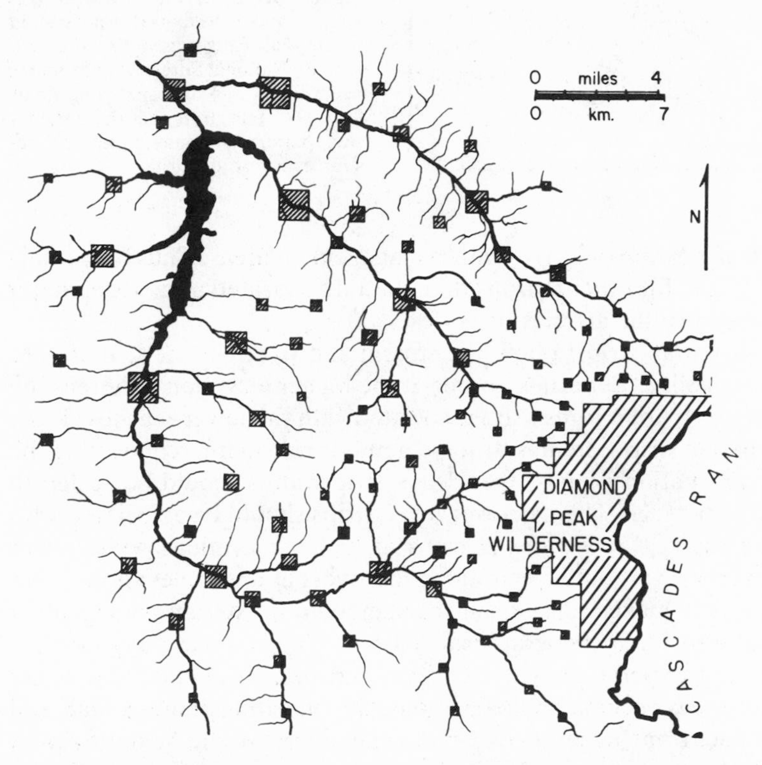

Figure 10.2 A possible spatial and size-frequency distribution of different-sized old-growth islands along riparian strips at progressively greater distances from a present wilderness area in the Willamette National Forest. 3 × 640 acres (260 ha), 7 × 320 acres (130 ha), 15 × 160 acres (65 ha), 30 × 80 acres (32 ha), 60 × 40 acres (16 ha) equal to 5% of the total area.

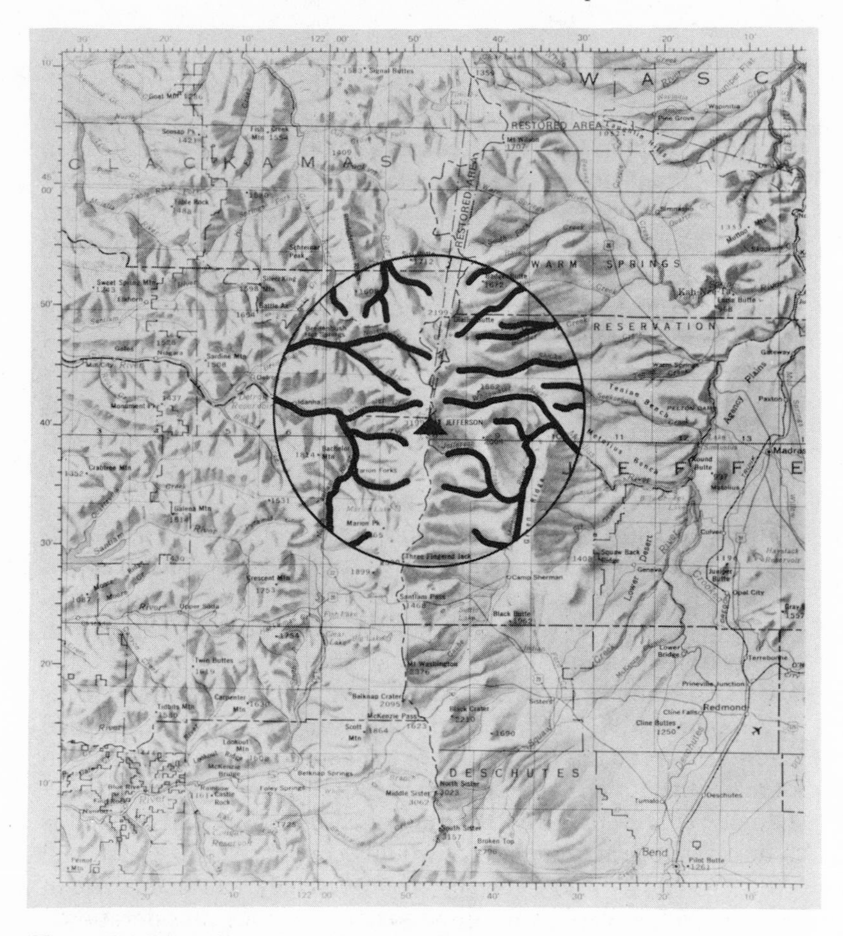

Figure 10.3 Rivers from one drainage are frequently juxtaposed with rivers from a different drainage. Protected riparian strips along these rivers facilitate animal movement and gene flow between populations.

are juxtaposed on opposite sides of the Cascade Range (fig. 10.3). Every opportunity should be taken to position islands near these juxtaposed streams to enhance the connectivity and functionality of the habitat island system.

It should be noted that using the natural dendritic patterns of the landscape as a skeleton for the system of islands will necessitate alterations in human use patterns. In particular, some roads might have to be closed, since the same dendritic pattern, as pointed out earlier, provides the major pathways for human use.

Because the habitat island system is being designed to last centuries into the future, present occurrence of old growth on a site is an ideal but should not be considered either a necessary or a sufficient rationale for choice. While the probability of destruction by natural catastrophe may be trivial in any given year, the combined probability of destruction in a period of 200 years might approach 100%. This implies that strong consideration must be given to the concept of replacement stands (implicit in the long-rotation island strategy) and the issue of forest protection. The probability of destruction is directly related to the frequency of occurrence of destructive agents such as fire, windstorm, insect and disease epidemic, landslide, and snowslide, and inversely related to the effectiveness of prevention. The emphasis here should be on fire.

Historically, fire has burned approximately 0.24% of the Western Cascades forest acreage annually (Andrews and Cowlin 1940), leading to an early observation that about 18% of the timbered area had been "recently" burned (Langille et al. 1903). This led to observations such as "the most startling feature shown by the land-classification map of this state is the extent of the burned areas . . ." (Gannett 1902, 11). The cumulative impact was also described as follows: "probably 90 percent of the entire area examined has at some remote period suffered from fires of which traces still remain" (Langille et al. 1903, 88); "in estimating future fire depletion it was assumed that a catastrophic fire would occur once in 30 years . . ." (Andrews and Cowlin 1940, 50). The frequency, cause, and size of fire, and people's ability to suppress it are obvious considerations.

Fire does not have an equal probability of occurrence on all sites, and the likelihood of a stand's total destruction is certainly influenced by fuel load, the rate of spread, and fire management possibilities. Whereas the frequency and extent of lightning fires is much greater at higher elevations, human-caused fires are associated with land use and tend to follow transportation routes and activity centers such as campsites (Burke 1979). Because forest management operations also constitute an important source of fire, access to the vicinity of old-growth areas should be discouraged. Analysis of fires that occurred just prior to 1901 reveals that whereas 13.4% of the acreage above 4,000 feet (1,220 m) was burned, only 3.4% of the acreage below 4,000 feet (1,220 m) had recently burned (fig. 10.4). Similarly, maps of human-caused fires in the

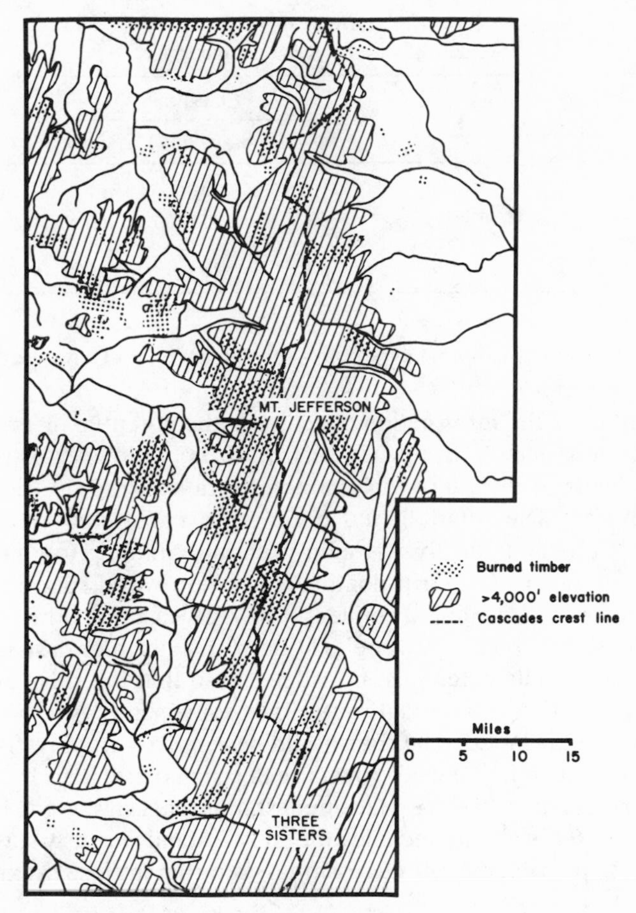

Figure 10.4 Distribution of "recently" burned timber as mapped in 1903 by Langille and his associates in relation to elevation. The proportion of acreage above 4,000 feet (1,200 m) that was recently burned is much higher ($P \leq 0.001$) than the proportion of acreage below 4,000 feet that was burned.

central portion of the Willamette National Forest reveal a striking association of fire frequency and human activity (Burke 1979). When factors such as fuel load and spread rate are also considered, it is possible to draw conclusions about which sites are most prone to destruction and which are most secure. In table 10.1 I have ranked the security of sites by elevation and aspect, with 1 being most secure and 12 being least secure. While low- to medium-elevation sites on north and east slopes are most secure, high-elevation sites

TABLE 10.1
Security of sites by elevation and aspect.

		Elevation		
		low	medium	high
Aspect	North	1	2	5
	East	3	4	6
	West	7	9	11
	South	8	10	12

with a south or southwest aspect are least secure (Al Lang, personal communication).

Combining the information concerning the potential diversity of vertebrate species with that on security from various catastrophic forces leads to the following characterization of the priority old-growth site. The stand should occur on a moist site containing surface water and, ideally, a stream. It should contain a topographic bench and a riparian strip dominated by hardwood species. This same riparian strip should connect it with at least one other stand. The site should be at a lower elevation with a north or east aspect, but would ideally extend over a ridge top so that the ridge system could be used as a dispersal route; thus some sunny, south-facing area will be included. The site should be removed from traffic and the high probability of wild crown fire deriving therefrom. It should be nearly surrounded by replacement stands that can serve as buffer areas, but these should include at least two stands in early stages of growth to provide the full successional spectrum in close proximity.

11

Summary and Characteristics of the Island Archipelago Approach

Strategies for maintaining biotic diversity and minimum viable populations of wildlife on public forest lands have become increasingly reliant on planning and decision-making approaches. A sufficient amount of previously uncut forest remains west of the crest of the Cascades Mountains to establish a system of old-growth habitat islands. Principles of island biogeography can guide the planning of such a system.

The purpose of this work has been to evaluate the existing states of nature and the applicability of island biogeography principles, and to develop the most promising management alternative. The consequences that derive from applying this approach in the Douglas fir region will be shown to have widespread use wherever the old-growth or habitat island issue exists. The purpose of this chapter is to recount background information and premises used to develop the approach, and then identify the specific properties and characteristics of the approach that will have utility elsewhere.

Over 25 million acres (10 million ha) of Douglas fir–western hemlock forest originally covered an area 500 miles by 100 miles (800 × 160 km) on the western slopes of the Cascade Mountains of Washington and Oregon. The combination of high latitude (42–49° N latitude) and rugged terrain greatly accentuates the importance of aspect and elevation in creating a diverse landscape mosaic of habitat types. Conifer species richness is greater than elsewhere in North America and hardwood species are very few; thus conifer dominance is very pronounced. The hardwoods are mostly limited to riparian strips, wet sites, and low elevations. Since only 10% of the annual rainfall occurs during the higher temperature and reproductive season months, streams and standing surface water are critical to wildlife. The association of water and hardwoods gives the riparian habitat type added value. The massive, long-lived nature of the conifers and equally large snags and logs contribute temporal stability to the system. The broken canopy and uneven age and size distribution of trees in old-growth stands allow the distribution of foliage throughout the vertical profile and a patchy horizontal distribution near the forest floor. Numerous snags and broken-top trees provide roosting and nesting sites for birds and some mammals, and a medium for wood-burrowing beetles and other arthropods that are the food base for many species of wildlife. Foliose, epiphytic lichens in the canopy and hypogeous mycorrhizal fungi in the duff are important as the bases of other food chains.

More families of breeding birds occur in the region than in any other area in North America and the density of species of birds and mammals is much greater than in most other regions. The mature and old-growth forests and attendant riparian strips of western Oregon support the only two folivorous (red tree vole and white-footed vole) and the only fungivorous (California red-backed vole) mammal species to occur in North America. Perhaps because of the low numbers of hardwood species, the northern flying squirrel relies heavily on the foliose lichens and is the only arboreal lichen consumer in North America. The number of carnivorous species is high and carnivores constitute 65% of the vertebrate fauna. This preponderance of carnivores is possible only because of the diverse array and productivity of food chains that support them.

Lower-elevation sites support many more species than higher elevations, and moist sites support more than very wet or dry sites. Mature and old-growth forests provide primary habitat for 118 species of vertebrates, of which more than one-third do not find primary habitat outside this forest type. Seedling-sapling stages of succession support more species than old growth, but many of these are common species in several other habitats. About forty-five species will not occur in second-generation short-rotation forests that do not contain abundant snags, broken-top trees, and fallen logs.

The trends and patterns of cutting the old-growth forest have caused widespread concern. For nearly fifty years annual loss and removal of timber have greatly exceeded annual growth. Old growth has been virtually eliminated from privately owned lands and lower-elevation sites. Old growth from public lands is expected to provide the major timber supply in the immediate future but for western Oregon as a whole there is a projected 22% reduction in cut by the year 2000. A severely skewed age distribution of existing stands of timber portends reductions in the abundance of game species as well as nongame old-growth wildlife as old growth is further reduced and present regeneration acreage reaches mid-rotation age. Although old growth Douglas fir still occurs on about 25% of national forest lands west of the Cascades, it has been reduced to only 3.3% of the Siuslaw National Forest. Remaining patches of old growth are small and isolated.

Island biogeography theory predicts that remnant patches of old growth salvaged from much larger, continuous stands will support greater numbers of species than will comparably sized replacement stands developed in isolation from contiguous old-growth forests. This prediction provides incentive to conserve existing old growth rather than depend on the development of replacement stands.

From the long-term perspective, the absence of immigration has caused a progressive depletion of the boreal mammal faunas of numerous Great Basin montane habitat islands. In a shorter time frame, the mammal fauna of Mount Rainier National Park has been reduced by about thirteen species in the last sixty years. This empirical evidence indicates that old-growth islands will not maintain their present species richness unless we ensure a colonization rate equal to known extinction rates. This equalization of local extinction and colonization rates can only be achieved by facilitating the movement of animals between islands. Larger habitat islands will initially support more species than smaller ones and the loss rate will not be as great from large islands. On the other hand, for any given total acreage commitment, it is not clear that larger but fewer islands will maintain more species. In the case of the western Cascades just the opposite seems to be true. Because a large proportion of the vertebrate species are wide-ranging carnivores, even the largest existing tracts of old-growth cannot be expected to "contain" complete faunal communities. The majority of species most seriously threatened by extinction are wide-ranging and will utilize landscapes sustantially disrupted by production-oriented systems. A well-integrated old-growth island system consisting of a large number of islands interspersed throughout the matrix of managed forest is probably a better alternative for such species. Added reasons for this approach are that the number of possible islands increases geometrically as average size is decreased. Average distance to the nearest neighbor island is reduced geometrically as the number is increased. This means that if total area is held constant, small reductions in average island size greatly increase the number of islands possible and greatly reduce the travel distance between islands.

Biotic diversity must be recognized as much more than species diversity. Critical components of genetic variability within species must be recognized and managed for at the same time that we strive to maintain the integrity of coevolved communities. Species characterized by small body size, habitat specificity, and sedentary habits are likely to manifest a high degree of clinal and/or ecotypic variation. Numerous replicate habitat patches should be allocated to maintain this isozyme diversity. For any given number of individuals, species characterized by large, wide-ranging animals with less specific habitat requirements will manifest lower isozyme diversity within the species. But in this case, the more numerous and widespread distribution of habitat islands will be strategic in maintaining the species itself. The full spectrum of genetic resources can only be conserved by focusing on the heterogeneous landscape mosaic. A system of islands that facilitates the development of locally adapted ecotypes while not preventing natural levels of gene flow between demes is ideal. If a species should fall to critically low population levels, it is important to facilitate its rapid buildup in population size to prevent bottlenecking and undue inbreeding.

Three factors influence effective island size: actual size, distance between islands, and hostility of the environmental matrix between islands. Because of climatic effects (e.g., wind penetration), threat of fire, dissimilarity of habitat types from the standpoint of vertebrate species, and other factors, an old-growth island surrounded by clearcut should be several times larger than an island surrounded by mature timber. Thus effective island size can be increased by regulating the composition of surrounding timber stands. Long rotation management areas surrounding each old-growth island will provide numerous advantages and are judged superior to the alternative of allowing short-rotation plantations to abut the old-growth islands.

An integrated system of long-rotation islands must consider the nature of the islands themselves, their specific location, their size-frequency distribution and spatial arrangement, and the connectivity and physical linkages between islands. Defining a long-rotation period as equal to four short-rotation periods will ensure that any given stand within the long-rotation unit will be classified as "old growth" ($\geq$ 240 years) for 25% of its total rotation length. At equilibrium, 25% of the total acreage committed to long-rotation management will occur as old growth. Surrounding each old-growth core area with nine replacement stands in various stages of growth would ensure that roughly 10% of the surrounding area would be in a regeneration stage, 30% would be less than 100 years old, 50% would be mature timber between 100 and 250 years of age, and 10% of the surrounding area would be old growth.

Deciding the average size of old-growth islands is no more or less important than deciding the nature of the size frequency distribution of the old-growth islands. The log-normal distribution is common in ecology and is suggested for use. This implies that a large number of smaller stepping-stone islands can be used to link the fewer but larger and more distant old-growth islands to the present system of wilderness areas and preserves. The dendritic pattern of streams, riparian strips, roads, and related resource distribution should be borne in mind when choosing old-growth set-asides. Large areas should be placed near higher-order streams that are a considerable distance from the present high-elevation preserve system. Medium-sized areas can be associated with streams of second or third order that occur at medium distances from the preserves. The numerous small areas should be strategically located along lower-order (i.e., smaller streams) and serve as stepping stones for colonizers moving to and from the preserves. All should be interconnected by riparian strips or comparable travel corridors.

All factors considered, an ideal old-growth area would occur on a moist site containing surface water and a stream. It would contain a topographic bench and a riparian strip dominated by hardwood species. This same riparian strip would connect it with at least two other stands. The site would be at a low elevation with a north or east aspect, but would ideally extend over a ridge top so that the ridge system could be used as a dispersal route and so some sunny, south-facing area would be included. The site would be as far removed from traffic and attendant risks such as wild crown fire as possible. It would be nearly surrounded by replacement stands that can serve as a buffer area and by at least two stands in early stages of growth to provide the full spectrum of successional stages in close proximity. While old growth refers to a community and/or ecosystem type, old age refers to a single chronological parameter. Until such a time as the two are shown to be synonymous or until it is demonstrated that we can recreate old growth, emphasis should be focused on conservation of existing old growth. Secondary emphasis should be placed on generation of strategically located replacement stands.

Specific properties of the island archipelago approach are specified below.

1. The first and foremost characteristic of the approach is the integration of conservation planning with development planning. I consider it imprudent to perpetuate the notion that conservation interests can be relegated to one corner of the world while developmental interests are relegated to another. This pattern of thought is especially threatening when it leads to complacency and the illusion that we do not have to be responsible for our actions in one area because we have made our sacrifice in another. A small number of discrete areas would be no more effective at preserving biotic diversity than a set of clean air or clean water reserves would be at conserving those resources. Some species of wildlife are highly mobile. An "extensive" approach founded on principles of landscape and regional ecology must be developed to complement the "intensive" approach represented by national parks and wildlife refuges. This extensive approach should be based on sound principles of multiple use, not preemptive policies of exclusive use. The archipelago approach shifts the emphasis away from any single old-growth habitat island toward a system of islands integrated into the managed forest landscape.

2. The island archipelago approach hinges on the notion that there are different, compelling, and noncompromisable reasons for committing specific habitat islands to the maintenance of biotic diversity. These reasons should be viewed not as competitive but as complementary. Some habitat islands will be chosen for geographic reasons, some for their intrinsic diversity, some for a certain species, some for their endemics, some for their contribution to within-species genetic diversity, and still others because of their contribution to the archipelago system. Indeed, the overriding rationale for selection of any given habitat island should be its contribution to the overall system. With reference to the application at hand, this approach shifts emphasis from the old-growth ecosystem to the landscape system of old-growth islands.

3. If maintenance of biotic diversity is the objective, then habitat islands should not be chosen because of their present role as much as for the role they will play in the future. In the human-dominated landscape of the future, many species will not have a "continental" source pool of potential colonizers other than those maintained

within the system of preserves and/or habitat islands (for example, North American grizzly bears or large herbivores in African parks). This means that the degree to which the habitat island system fuctions in a multiple-use environment will determine survival or extinction for a considerable number of species. Based on the combined body of wildlife biology information and general ecological principles, I believe that an island system chosen by design will be superior to one inherited by default. Design criteria should be based on contribution to, and anticipated function of, the archipelago system rather than simple evaluation of individual islands.

4. Like any system, the habitat island archipelago will consist of components that are expected to function in a certain manner, in this case to maintain biotic diversity. These components will have identifiable characteristics including a size distribution and a space distribution, and it is hoped these two distributions will be interdependent. The individual islands will occur in a specific location, and to simplify planning, the locations should be permanent rather than shifting. Whether by design or otherwise, animal movements will cause the component islands to interact. This interaction is essential to the perpetuation of within-species genetic diversity and the survival of many species. A system of interconnecting travel corridors should therefore be designed into the archipelago system.

5. As eloquently described by Dubos (1970, 1973), different sites have differential value for the various uses to which we commit them. Site capability and "genius" are concepts well known to foresters and agriculturalists. Some sites are clearly more appropriate for timber production while others are more easily adapted to recreation, and yet others will serve best for maintenance of biotic diversity. The concept of multiple use is based on the notion that when sites and uses are chosen in certain combinations they will interact synergistically, while if chosen inappropriately, antagonism may result. The placing of clearcuts serves as a ready example. Several clearcuts scheduled adjacent to one another in close sequence may cause excess runoff and nutrient loss, negative aesthetic and wildlife population responses, and in the extreme, slumping or landsliding and destruction of the site itself. Clearcuts located in other shapes, sizes, and patterns may generate positive effects multiplicatively. The edge effect is an example. In designing strategies and mosaics for the maintenance of biotic diversity, it is important to bear the concept of ecosystem synergism in mind.

Old-growth stands in close proximity to surface water and hardwood riparian forests, and interconnected to other islands in the archipelago will function better than if the contextual considerations are overlooked. This is in part because these different ecosystems will function asynchronously, some producing maximum habitat value at one time while others are utilized at a different time. Some habitat types are essential for one life stage (for example, amphibians that have larval and adult life stages) while a totally different habitat type is required for a different life stage. Any given habitat island will produce excess propagules in one year but perhaps not in several others. Opportunity for these time-space interactions must be designed into the functioning landscape system. The juxtaposition of ecosystem types and the linkage of islands so that they function as an archipelago are as important as the conservation of old growth itself.

6. In order to maximize the effective island size, provide a reasonable predictability of location from one decade to another, and provide a spatial gradation from the old-growth core area into the young-growth production forest, each old-growth island should be surrounded by a long-rotation management area. While the physical properties of the long-rotation buffer zone will produce tangible benefits, the graduated shift in purpose from conservation areas to high-production areas is equally significant. The short-rotation plantations have timber production as a primary purpose and the provision of other multiple-use benefits as subsidiary objectives. Among the objectives of the old-growth core areas, the maintenance of biotic diversity is the highest. Long-rotation management areas lie both physically and conceptually between the two, with the full spectrum of multiple-use benefits on a more equal ranking.

Generalizing from old-growth issue to conservation strategy at large yields the concept of multiple-use modules (MUMs). Each MUM consists of a medullary area with a preservation objective that is surrounded by two or more cortex areas with differing multiple-use objectives. The whole module should be programmed into the context of a production-oriented landscape (fig. 11.1). This allows the preservation areas to be buffered from the harsh impacts and vicissitudes of the human-dominated landscape while also mitigating the depredation of sometimes hostile wildlife species (alligators and panthers in Florida, bears and wolves in western states, elephants and leopards in Africa) on humans and their possessions. I envision that vehicles, roads, and consumptive uses would be

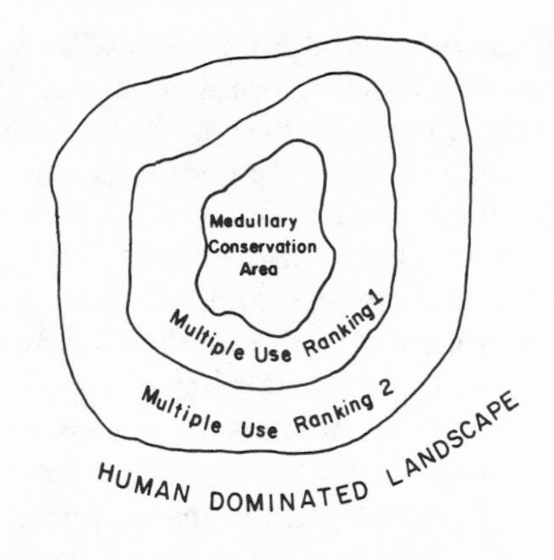

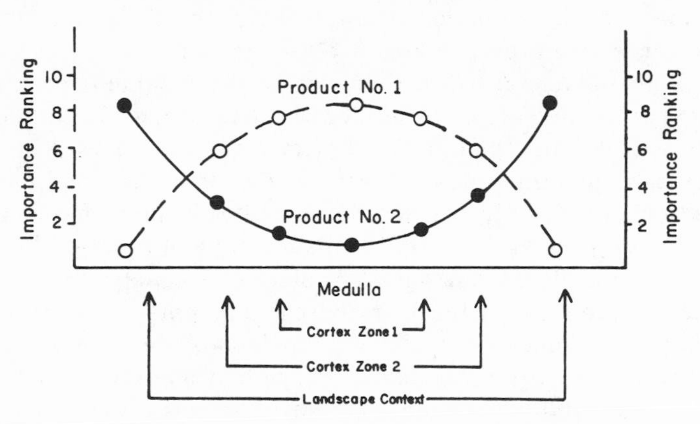

Figure 11.1 Conceptual scheme of how core conservation areas might be surrounded by cortical layers of increasingly intensive management. Progression outward from the core area into the human-dominated landscape would lead to an inverse relation between product valuations as priorities change. Multiple-use modules (MUMs) such as these should facilitate the integration of conservation planning with development planning.

disallowed in the medullary area; that perhaps roads and consumptive uses would be authorized in the one or more cortex zones; and that large depredating wildlife species would be strongly discouraged from the human-dominated landscape matrix.

There are corollaries to this concept. Multiple-use management requires that several products be given consideration, but it does not specify that the products be considered equally. This concept allows for an inverse relation between relative importance of several products and the management zone being considered (fig. 11.1). As one product (e.g., biotic diversity) becomes relatively less important as we move outward from the core area, a second product (e.g., fiber production) becomes relatively more important. Thus the importance ranking of the various objectives is dependent on the distance removed from the core area (fig. 11.1).

A second corollary is that the size and degree of protection afforded the core area should also be directly proportional to the intensity with which the human-dominated landscape is managed and inversely proportional to the amount of surrounding buffer area. If low-intensity forest management (e.g., long rotation and select cutting) were to dominate the landscape matrix, then requirements for natural area set-asides would be much less than if the landscape consisted of tree plantations or agronomic crops. If it were possible to manage the entire forest landscape in a very low-intensity, long-rotation manner, there would be little if any need for special provision areas.

7. Potential energy exists whenever a differential in the concentration or location of like molecules can be identified. Water molecules at 5,000 feet elevation and subject to gravitational pull have a measurable potential energy. Earth and rock uplifted to high elevations by geophysical forces, and waves created by lunar and centrifugal forces, possess massive amounts of potential energy. The low concentration of ions of salt in freshwater rivers juxtaposed with the higher concentration in the sea represents a potential energy source similar to the ionic differential in a dipolar storage battery. All of these are nonsolar sources, and although they all exist in the landscape, their utility to organisms and ecosystems remains to be demonstrated by ecologists of the future. Suffice it to say that water rushing downslope from higher to lower elevations, both carrying and depositing inorganic and organic nutrients, represents a kinetic energy source important to landscape ecologists. Landscapes with great topographic relief transfer most of this energy via a system of dendritic tributaries, main channels, and distributaries. These energy pathways dissect the landscape and in many cases leave energy signatures that may be interpreted (fig. 11.2). The "drainage density" (i.e., number per unit area) and degree to which these signatures are pronounced is a direct function of the intensity of energy flow, or in this case, the steepness of topographic relief. The physical form of the landscape and the energy signatures, in turn, constitute a landscape template that should be used when designing resource management systems. Watersheds of different order should constitute a basic resource mapping and management unit

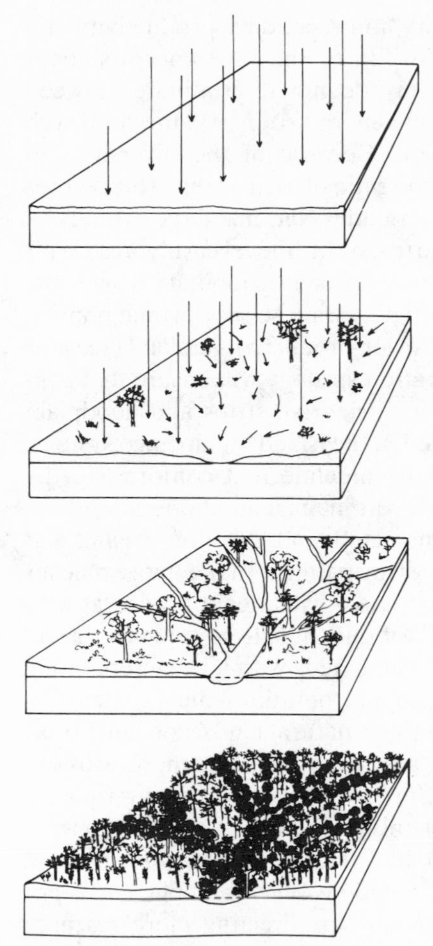

Figure 11.2 Landscapes are subjected to numerous nonsolar auxiliary energy sources. As the solar and nonsolar energy is transformed (it cannot be destroyed), it does work such as degrading, aggrading, and transporting. Some ecosystems benefit from these processes and are therefore richer in structure, resource diversity, and productivity than others. Strategies aimed at conservation of biotic diversity must recognize and build on these principles.

just as streams of different order will constitute the connectors between units. Conservation strategies should be ecologically designed to fit the landscape rather than architectually designed for aesthetic appeal.

Interpretations of these basic patterns might develop along the following lines. Although greater amounts of potential energy may be associated with high-elevation upland systems, this energy is generally more diffuse and less utilizable. The generally more usable forms of energy occur in the concentrated pathways (streams and riparian strips) and/or lower-elevation floodplain sites. These sites will manifest higher primary and secondary productivity and will support higher densities of wildlife. Since the energy sources are diverse (earthworms probably do not discriminate between allochthonous and autochthonous energy sources), the food web will be correspondingly complex. Because of the diversity and productivity of the food chains, the ratio of carnivores to herbivores will generally increase (in the Everglades National Park virtually all of the vertebrates are third-, fourth-, or fifth-level carnivores). This means that for any given minimum effective population size, the area of high-productivity lowland necessary to support the population will be less than the area of lower-productivity upland (see also Foster 1980). The ideal conservation strategy will ensure that concentrated energy pathways such as riparian strips and floodplain forests will be heavily represented in the island archipelago system.

A second example of a design guideline that conforms to the natural landscape is directed at forest management operations and impact mitigation. In a highly dissected landscape the streams and rivers will form an elaborate dendritic pattern. In the upper reaches of the system the intervening land areas will appear triangular with an apex pointed downslope. Half of this triangle of upland area will derive from each of two paired watersheds with a ridge separating the two. Perhaps forest management operations such as clearcutting should be scheduled within these natural landscape units (i.e., straddling the ridge). If so, the adverse effects of runoff, erosion, and nutrient loss would be buffered by the two flanking riparian strips and streams as opposed to only one stream if the activity were contained within a single watershed (Harris and Marion 1981).

Many additional applications could be developed but that is not the point to be made here. Rather, by directing more research toward the analysis and interpretation of auxiliary energy sources and landscape patterns, we will be in a better position to design with nature rather than defend against her.

8. The island archipelago approach has been developed as a forest management strategy. Thus, even though the same concepts and recommendations may be widely applicable, certain reservations must also be entered into the record. One of the most compelling arguments for the archipelago approach is that set-aside areas sufficiently large to protect the wide-ranging species simply cannot be achieved. Therefore, the attitudinal and resource commitment must be reallocated from the "intensive" park and preserve approach to the more dispersed, "extensive" approach. But this

same argument does not hold for park and refuge systems unless a migratory species or group (for example the waterfowl refuge system) is being considered. If *in situ* conservaticn of large resident species in parks and preserves is the objective, then each of these areas must be sufficiently large to totally contain effective populations of the designated species. Road kills, accidental shootings, and otherwise fatal encounters with humans will probably remain the principal mortality factors for large depredating species. This means that the conservation of these species wil be dependent upon minimizing the number of negative encounters between them and humans. Thus, the intensive approach, where large parks and preserves are designated for large animal conservation, must be integrated with the extensive multiple-use management approach. They are in no way competitive, nor can either approach successfully maintain existing levels of biotic diversity. Multiagency cooperation would seem to be the next logical step in the development of a comprehensive, effective conservation program.

Appendix 1

Scientific names of species cited in text. Names for plant species from Little (1979), amphibians and reptiles from Stebbins (1966), birds from Gruson (1976), and mammals from Maser et al. (1981).

Flora

sugar pine	*(Pinus lambertiana)*
western white pine	*(Pinus monticola)*
whitebark pine	*(Pinus albicaulis)*
longleaf pine	*(Pinus palustris)*
slash pine	*(Pinus elliottii)*
ponderosa pine	*(Pinus ponderosa)*
lodgepole pine	*(Pinus contorta)*
western larch	*(Larix occidentalis)*
Engelmann spruce	*(Picea engelmannii)*
Douglas fir	*(Pseudotsuga menziesii)*
western hemlock	*(Tsuga heterophyla)*
mountain hemlock	*(Tsuga mertensiana)*
subalpine fir	*(Abies lasiocarpa)*
grand fir	*(Abies grandis)*
Pacific silver fir	*(Abies amabilis)*
noble fir	*(Abies procera)*
incense cedar	*(Libocedrus decurrens)*
western red cedar	*(Thuja plicata)*
Alaska cedar	*(Chamaecyparis nootkatensis)*
Pacific yew	*(Taxus brevifolia)*
red alder	*(Alnus rubra)*
Gary oak	*(Quercus garryana)*

Fauna

Amphibians

northwestern salamander	*(Ambystoma gracile)*
long-toed salamander	*(Ambystoma macrodactylum)*

Pacific giant salamander	*(Dicamptodon ensatus)*
Olympic salamander	*(Rhyacotriton olympicus)*
Rough-skinned newt	*(Taricha granulosa)*
Dunn's salamander	*(Plethodon dunni)*
western red-backed salamander	*(Plethodon vehiculum)*
Larch mountain salamander	*(Plethodon larselli)*
Del Norte salamander	*(Plethodon elongatus)*
Siskiyou mountain salamander	*(Plethodon stormi)*
Oregon salamander	*(Ensatina eschscholtzi)*
Oregon slender salamander	*(Batrachoseps wrighti)*
California slender salamander	*(Batrachoseps attenuatus)*
black salamander	*(Aneides flavipunctatus)*
clouded salamander	*(Aneides ferreus)*
tailed frog	*(Ascaphus truei)*
western toad	*(Bufo boreas)*
Pacific treefrog	*(Hyla regilla)*
red-legged frog	*(Rana aurora)*
spotted frog	*(Rana pretiosa)*
Cascades frog	*(Rana cascadae)*
Foothill yellow-legged frog	*(Rana boylei)*

Reptiles

alligator	*(Alligator mississipiensis)*
western pond turtle	*(Clemmys marmorata)*
painted turtle	*(Chrysemys picta)*
western fence lizard	*(Sceloporus occidentalis)*
western skink	*(Eumeces skiltonianus)*
southern alligator lizard	*(Gerrhonotus multicarinatus)*
northern alligator lizard	*(Gerrhonotus coeruleus)*
rubber boa	*(Charina bottae)*
ringneck snake	*(Diadophis punctatus)*
sharp-tailed snake	*(Contia tenuis)*
western yellow-bellied racer	*(Coluber constrictor mormon)*
gopher snake	*(Pituophis melanoleucus)*
common kingsnake	*(Lampropeltis getulus)*
California mountain kingsnake	*(Lampropeltis zonata)*
common garter snake	*(Thamnophis sirtalis)*
western terrestrial garter snake	*(Thamnophis elegans)*
western aquatic garter snake	*(Thamnophis couchi)*
northwestern garter snake	*(Thamnophis ordinoides)*
western rattlesnake	*(Crotalus viridis)*

Birds

California condor	*(Cymnogyps californianus)*
osprey	*(Pandion haliateus)*
bald eagle	*(Haliaeetus leucocephalus)*
sharp-shinned hawk	*(Accipiter striatus)*
Cooper's hawk	*(Accipiter cooperii)*
goshawk	*(Accipiter gentilis)*
red-tailed hawk	*(Buteo jamaicensis)*
merlin	*(Falco columbarius)*
blue grouse	*(Dendrogapus obsurus)*
ruffed grouse	*(Bonassa umbellus)*
heath hen	*(Tympanuchus cupido)*
bobwhite	*(Colinus virginianus)*
scaled quail	*(Callipepla squamata)*
mountain quail	*(Oreortyx pictus)*
valley quail	*(Laphortyx californicus)*
Japanese quail	*(Coturnix japonica)*
ring-necked pheasant	*(Phasianus colchicus)*
chukar partridge	*(Alectoris chukar)*
Hungarian partridge	*(Perdix perdix)*
whooping crane	*(Grus americana)*
rock dove	*(Columba livia)*
band-tailed pigeon	*(Columba fasciata)*
mourning dove	*(Zenaida macroura)*
passenger pigeon	*(Ectopistes migratorius)*
screech owl	*(Otus asio)*
great horned owl	*(Bubo virginianus)*
pygmy owl	*(Glaucidium gnoma)*
spotted owl	*(Strix occidentalis)*
long-eared owl	*(Asio otus)*
saw-whet owl	*(Aegolius acadicus)*
night hawk	*(Chordeiles minor)*
Vaux's swift	*(Chaetura vauxi)*
rufous hummingbird	*(Selasphorus rufus)*
Calliope hummingbird	*(Stellula calliope)*
flicker	*(Colaptes auratus)*
pileated woodpecker	*(Dryocopus pileatus)*
Williamson's sapsucker	*(Sphyrapicus thyroideus)*
yellow-bellied sapsucker	*(Sphyrapicus varius)*
hairy woodpecker	*(Picoides villosus)*
downy woodpecker	*(Picoides pubescens)*

northern three-toed woodpecker	*(Picoides tridactylus)*
olive-sided flycatcher	*(Contopus borealis)*
western wood pewee	*(Contopus sordidulus)*
Traill's flycatcher	*(Empidonax traillii)*
Hammond's flycatcher	*(Empidonax hammondii)*
western flycatcher	*(Empidonax difficilis)*
dusky flycatcher	*(Empidonax oberholseri)*
purple martin	*(Progne subis)*
tree swallow	*(Tachycineta bicolor)*
gray jay	*(Perisoreus canadensis)*
Stellar's jay	*(Cyanocitta stelleri)*
scrub jay	*(Aphelocoma coerulescens)*
crow	*(Corvus brachyrhynchos)*
black-capped chickadee	*(Parus atricapillus)*
chestnut-backed chickadee	*(Parus rufescens)*
common bushtit	*(Psaltriparus minimus)*
white-breasted nuthatch	*(Sitta carolinensis)*
red-breasted nuthatch	*(Sitta canadensis)*
brown creeper	*(Certhia familiaris)*
Bewick's wren	*(Thryomanes bewickii)*
winter wren	*(Troglodytes troglodytes)*
golden-crowned kinglet	*(Regulus satrapa)*
ruby-crowned kinglet	*(Regulus calendula)*
western bluebird	*(Sialia mexicana)*
mountain bluebird	*(Sialia currucoides)*
Townsend's solitaire	*(Myadestes townsendi)*
Swainson's thrush	*(Catharus ustalatus)*
hermit thrush	*(Catharus guttatus)*
robin	*(Turdus migratorius)*
varied thrush	*(Zoothera naevica)*
European starling	*(Sturnus vulgaris)*
solitary vireo	*(Vireo solitarius)*
Hutton's vireo	*(Vireo huttoni)*
warbling vireo	*(Vireo gilvus)*
orange-crowned warbler	*(Vermivora celata)*
Nashville warbler	*(Vermivora ruficapilla)*
yellow warbler	*(Dendroica petechia)*
Wilson's warbler	*(Wilsonia pusilla)*
yellow-rumped warbler	*(Dendroica coronata)*
black-throated gray warbler	*(Dendroica nigrescens)*
Townsend's warbler	*(Dendroica townsendi)*

hermit warbler	*(Dendroica occidentalis)*
MacGillivray's warbler	*(Geothlypis philadelphia)*
English sparrow	*(Passer domesticus)*
western tanager	*(Piranga ludoviciana)*
black-headed grosbeak	*(Pheucticus melanocephalus)*
evening grosbeak	*(Coccothraustes vespertina)*
lazuli bunting	*(Passerina amoena)*
Cassin's finch	*(Carpodacus cassinii)*
purple finch	*(Carpodacus purpureus)*
pine siskin	*(Carduelis pinus)*
rufous-sided towhee	*(Pipilo erythrophthalmus)*
American goldfinch	*(Carduelis tristis)*
red crossbill	*(Loxia curvirostra)*
savannah sparrow	*(Ammodramus sandwichensis)*
vesper sparrow	*(Pooecetes gramineus)*
Oregon junco	*(Junco hyemalis)*
chipping sparrow	*(Spizella passerina)*
white-crowned sparrow	*(Zonotrichia leucophrys)*
fox sparrow	*(Zonotrichia iliaca)*
song sparrow	*(Zonotrichia melodia)*

Mammals

Oppossum	*(Didelphis virginianus)*
wandering shrew	*(Sorex vagrans)*
dusky shrew	*(Sorex obscurus)*
Yaquina shrew	*(Sorex yaquinae)*
Pacific shrew	*(Sorex pacificus)*
northern water shrew	*(Sorex palustris)*
marsh shrew	*(Sorex bendirei)*
Trowbridge shrew	*(Sorex trowbridgei)*
shrew mole	*(Neurotrichus gibbsi)*
coast mole	*(Scapanus orarius)*
California myotis	*(Myotis californicus)*
little brown myotis	*(Myotis lucifugus)*
Yuma myotis	*(Myotis yumanensis)*
long-eared myotis	*(Myotis evotis)*
fringed myotis	*(Myotis thysanodes)*
long-legged myotis	*(Myotis volans)*
silver-haired bat	*(Lasionycteris noctivagans)*
big brown bat	*(Eptesicus fuscus)*
hoary bat	*(Lasiurus cinereus)*

western big-eared bat	*(Plecotus townsendi)*
pallid bat	*(Antrozous pallidus)*
pika	*(Ochotona princeps)*
snowshoe hare	*(Lepus americanus)*
brush rabbit	*(Sylvilagus bachmani)*
mountain beaver	*(Aplodontia rufa)*
yellow-pine chipmunk	*(Eutamias amoenus)*
Townsend chipmunk	*(Eutamias townsendi)*
Siskiyou chipmunk	*(Eutamias siskiyou)*
Beechey ground squirrel	*(Spermophilus beecheyi)*
mantled ground squirrel	*(Spermophilus lateralis)*
western gray squirrel	*(Sciurus griseus)*
chickaree	*(Tamiasciurus douglasi)*
northern flying squirrel	*(Glaucomys sabrinus)*
Mazama pocket gopher	*(Thomomys mazama)*
beaver	*(Castor canadensis)*
deer mouse	*(Peromyscus maniculatus)*
old-field mouse	*(Peromyscus polionotus)*
dusky-footed woodrat	*(Neotoma fuscipes)*
bushy-tailed woodrat	*(Neotoma cinerea)*
California red-backed vole	*(Clethrionomys californicus)*
white-footed vole	*(Arborimus albipes)*
red tree vole	*(Arborimus longicaudus)*
heather vole	*(Phenacomys intermedius)*
Townsend vole	*(Microtus townsendi)*
long-tailed vole	*(Microtus longicaudus)*
Oregon vole	*(Microtus oregoni)*
Richardson vole	*(Microtus richardsoni)*
muskrat	*(Ondatra zibethicus)*
Norway rat	*(Rattus norvegicus)*
house mouse	*(Mus musculus)*
Pacific jumping mouse	*(Zapus trinotatus)*
porcupine	*(Erethizon dorsatum)*
coyote	*(Canis latrans)*
gray wolf	*(Canis lupus)*
red wolf	*(Canis rufus)*
red fox	*(Vulpes vulpes)*
black bear	*(Ursus americanus)*
grizzly bear	*(Ursus horribilis)*
ringtail	*(Bassariscus astutus)*
raccoon	*(Procyon lotor)*

marten *(Martes americana)*
fisher *(Martes pennanti)*
short-tailed weasel *(Mustela erminea)*
long-tailed weasel *(Mustela frenata)*
mink *(Mustela vison)*
wolverine *(Gulo luscus)*
spotted skunk *(Spilogale putorius)*
striped skunk *(Mephitis mephitis)*
river otter *(Lutra canadensis)*
cougar *(Felis concolor)*
Florida panther *(Felis concolor coryii)*
leopard *(Panthera Pardus)*
lynx *(Lynx canadensis)*
bobcat *(Lynx rufus)*
northern elephant seal *(Mirounga angustirostris)*
southern elephant seal *(Mirounga leonina)*
elephant (African) *(Loxodonta africana)*
javelina *(Dicotyles tajacu)*
Pere David's deer *(Elaphurus davidianus)*
Roosevelt elk *(Cervus elaphus)*
black-tailed or mule deer *(Odocoileus hemionus)*
white-tailed deer *(Odocoileus virginianus)*
fallow deer *(Dama dama)*
moose *(Alces alces)*
barren-ground caribou *(Rangifer tarandus)*
pronghorn *(Antilocapra americana)*
bison *(Bison bison)*
European bison *(Bison bonansus)*
mountain goat *(Oreamnos americanus)*
bighorn sheep *(Ovis canadensis)*

Appendix 2

Volume of timber cut from the Willamette National Forest in western Oregon for the years 1905–81. Data from Paulson and Leavengood 1977 and subsequent annual updates.

Year	million board feet	Year	million board feet	Year	million board feet
1905	14	1931	34.7	1957	373.2
1906	—	1932	15.1	1958	477.1
1907	16.3	1933	32.7	1959	614.6
1908	—	1934	28.2	1960	511.5
1909	—	1935	37.0	1961	597.0
1910	6.9	1936	39.0	1962	809.0
1911	1.1	1937	33.1	1963	671.1
1912	—	1938	43.9	1964	275.8
1913	0.6	1939	60.0	1965	636.8
1914	—	1940	62.2	1966	701.8
1915	5.2	1941	95.8	1967	678.3
1916	1.7	1942	85.9	1968	650.4
1917	3.4	1943	93.5	1969	671.9
1918	—	1944	132.4	1970	746.2
1919	—	1945	134.5	1971	718.0
1920	—	1946	117.3	1972	741.4
1921	0.5	1947	153.5	1973	734.1
1922	15.4	1948	207.0	1974	742.6
1923	21.8	1949	155.6	1975	777.2
1924	29.4	1950	308.5	1976	829.5
1925	51.4	1951	268.3	1977	—
1926	47.4	1952	371.2	1978	786.0
1927	59.4	1953	413.4	1979	770.9
1928	68.0	1954	408.8	1980	788.4
1929	66.0	1955	381.3	1981	774.5
1930	57.3	1956	402.8		

Appendix 3

Approximate acreage cut from different elevational zones in the Willamette National Forest. Data from Total Resource Inventory [TRI] System, Forest Supervisor's Office, Eugene, Oregon.

Elevation (feet)	Prior 1940	1941–50	1951–60	1961–70	1971–80
less than 2,000	440	2,413	3,332	6,302	5,750
2,000–3,000	519	4,973	16,253	26,574	23,168
3,000–4,000	85	4,217	19,329	36,550	31,994
greater than 4,000	27	938	5,368	18,541	17,215

Appendix 4

Forty-five terrestrial vertebrate species that require cavities, standing snags, and fallen logs in order to meet primary habitat requirements. Bird data from Meslow and Wright (1975); other data from Chris Maser.

Require cavities and/or standing snags	Require cavities and/or standing snags and/or large down logs
saw-whet owl	All species listed in
pygmy owl	left column plus:
tree swallow	
purple martin	
western bluebird	
mountain bluebird	
flicker	
yellow-bellied sapsucker	winter wren
black-capped chickadee	clouded salamander
screech owl	Oregon salamander
pileated woodpecker	western fence lizard
northern three-toed woodpecker	western skink
hairy woodpecker	southern alligator lizard
downy woodpecker	sharp-tailed snake
chestnut-backed chickadee	dusky shrew
red-breasted nuthatch	Yaquina shrew
white-breasted nuthatch	Trowbridge shrew
brown creeper	shrew-mole
spotted owl	coast mole
Vaux's swift	snowshoe hare
California myotis	yellow-pine chipmunk
Yuma myotis	Siskiyou chipmunk
little brown myotis	California red-backed vole
fringed myotis	spotted skunk
long-legged myotis	lynx
silver-haired bat	bobcat
big brown bat	

Require cavities and/or standing snags	Require cavities and/or standing snags and/or large down logs
raccoon	
marten	
fisher	

Appendix 5

List of research natural areas and comparable preserves in western Oregon and western Washington indicating size and management authority. Only those areas that are predominantly forested and occur west of the Cascades ridge are included. Numbers refer to figure 7.1. Sarah Greene is gratefully acknowledged for this information.

Name of research natural area	Ownership[a]	Area (ha)
1 North Fork Nooksack RNA[b]	USFS	605
2 Skagit River Bald Eagle Sanctuary	TNC	121
3 Cypress Island Eagle Cliffs	DNR	63
4 Wolf Creek RNA	USFS	61
5 Long Creek RNA	USFS	259
6 Wildflower Acres Preserve	TNC	10
7 Lake Twenty-two RNA	USFS	320
8 Quinault RNA	USFS	594
9 Thompson Clover RNA	USFS	81
10 Meeks Table RNA	USFS	28
11 Butter Creek RNA	USFS	226
12 Cedar Flats RNA	USFS	275
13 Steamboat Mountain RNA	USFS	567
14 Goat Marsh RNA	USFS	484
15 Sister Rocks RNA	USFS	87
16 Thornton T. Munger RNA	USFS	478
17 Sandy River Gorge Preserve	TNC	150
18 Bull Run RNA	USFS	146
19 Mill Creek RNA	USFS	330
20 High Peak RNA	BLM	618
21 Neskowin Crest RNA	USFS	476
22 Saddleback Mountain RNA	BLM	55
23 Bagby RNA	USFS	227
24 The Butte	BLM	17
25 Little Sink RNA	BLM	32
26 Carolyn's Crown RNA	BLM	105
27 Middle Santiam RNA	USFS	463

Appendix 5 *(continued)*

Name of research natural area	Ownership[a]	Area (ha)
28 Flynn Creek RNA	USFS	271
29 Grass Mountain RNA	BLM	316
30 Maple Knoll RNA	FWS	40
31 Cogswell-Foster Preserve	TNC	36
32 Metolius RNA	USFS	581
33 Wildcat Mountain RNA	USFS	405
34 Mohawk RNA	BLM	119
35 Ollalie Ridge RNA	USFS	291
36 Fox Hollow RNA	BLM	65
37 Camas Swale RNA	BLM	113
38 Elk Meadows RNA	BLM	83
39 Pringle Falls RNA	USFS	470
40 Gold Lake Bog RNA	USFS	188
41 Limpy Rocky RNA	USFS	760
42 Beatty Creek RNA	BLM	65
43 Cherry Creek RNA	BLM	239
44 North Myrtle Creek RNA	BLM	97
45 Abbott Creek RNA	USFS	1,077
46 Port Orford Cedar RNA	USFS	454
47 Coquille Falls RNA	USFS	202
48 Blue Jay RNA	USFS	85
49 Goodlow Mountain RNA	USFS	510
50 Brewer Spruce RNA	BLM	85
51 Woodcock Bog RNA	BLM	45
52 Ashland RNA	USFS	570
53 Wheeler Creek RNA	USFS	135
54 Winchuck Slope NAP	SLB	78

a. USFS = U.S. Forest Service (U.S.D.A.)
BLM = Bureau of Land Management (U.S.D.I.)
FWS = U.S. Fish and Wildlife Service (U.S.D.I.)
SLB = State Land Board (OR)
DNR = Department of Natural Resources (WA)
TNC = The Nature Conservancy
b. RNA = Research natural area

Literature Cited

Adams, D. M. 1977. Impacts of national forest timber harvest scheduling policies on softwood, stumpage, lumber, and plywood markets: An econometric analysis. Res. Bull. 15. Corvallis: For. Res. Lab., Oregon State Univ.

Airola, D. A., and R. H. Barrett. 1981. Vertebrates of Blodgett Forest Research Station, El Dorado County, California—An annotated species list. Spec. Pub. 3267. Berkeley: Div. Agric. Sci., Univ. California.

Allard, R. W. 1960. Principles of plant breeding. New York: John Wiley and Sons.

Allen, A. 1982. Habitat suitability index models: Marten. U.S.D.I. Fish and Wildl. Serv., FWS/OBS-82/10-11.

Anderson, R. C., O. L. Loucks, and A. M. Swain. 1969. Herbaceous response to canopy cover, light intensity, and throughfall precipitation in coniferous forests. *Ecology* 50:255–63.

Anderson, S. H. 1970. Ecological relationships of birds in forests of western Oregon. Ph.D. diss. Oregon State Univ., Corvallis.

———. 1972. Seasonal variations in forest birds of western Oregon. *Northwest Sci.* 46:194–206.

Anderson, S. H., and C. S. Robbins. 1981. Habitat size and bird community management. *Trans. North Am. Wildl. Nat. Resour. Conf.* 46:511–20.

Andrews, H. J., and R. W. Cowlin. 1940. Forest resources of the Douglas-fir region. Washington, D.C.: U.S.D.A. Misc. Pub. 389.

Aney, W. W. 1967. Wildlife of the Willamette Basin, present status. Portland: Basin Invest. Sect., Oregon State Game Comm.

Anonymous. 1936. Forest type maps, state of Oregon. Portland: U.S.D.A. For. Serv., PNW For. Expt. Sta. Portland, Oregon.

Anonymous. 1942. History of the Willamette National Forest. Corvallis: Mimeo report, Oregon State Univ. Library, SD428/W5/A2.

Anonymous. 1960. A preliminary survey of fish and wildlife resources of northwestern California. Portland: U.S.D.I. Fish and Wildl. Serv.

Anonymous. 1966. California fish and wildlife plan. Sacramento: California Dept. Fish and Game. 5 vols.

Anonymous. 1972. Notes and comment. *The New Yorker*, May 13, pp. 29–30.
Arrhenius, O. 1921. Species and area. *J. Ecol.* 9:95–99.
———. 1922. On the relation between species and area—a reply. *Ecology* 4:90–91.
Azevedo, J., and D. L. Morgan. 1974. Fog precipitation in coastal California forests. *Ecology* 55:1135–41.
Bailey, V. 1936. The mammals and life zones of Oregon. North Am. Fauna 55, U.S.D.A. Bur. Biol. Surv.
Beck, W. M. 1962. Studies on the vertebrate ecology of a bottomland woods in the central Willamette Valley. M.S. thesis, Oregon State Univ., Corvallis.
Behler, J. L., and F. W. King. 1979. The Audubon Society field guide to North American reptiles and amphibians. New York: Chanticleer Press.
Bekele, E. 1980. Island biogeography and guidelines for the selection of conservation units for large mammals. Ph.D. diss., Univ. Michigan, Ann Arbor.
Bengtsson, B. O. 1978. Avoiding inbreeding: At what cost? *J. Theoretical Biol.* 73:439–44.
Beuter, J. H., K. N. Johnson, and H. L. Scheurman. 1976. Timber for Oregon's tomorrow, an analysis of reasonably possible occurrences. Res. Bull. 19. Corvallis: For. Res. Lab., Oregon State Univ.
Bingham, C. W. 1976. North America's role in future wood supply. *J. Forest.* 74:512–14.
Blake, I. H. 1926. A comparison of the animal communities of coniferous and deciduous forests. *Illinois Biol. Monogr.* 10:371–521.
Bonnell, M. L., and R. K. Selander. 1974. Elephant seals: genetic variation and near extinction. *Science* 184:908–9.
Boucher, D. H. 1979. Seed predation and dispersal by mammals in a tropical dry forest. Ph.D. diss., Univ. Michigan, Ann Arbor. 216pp.
Bowles, J. B. 1963. Ornithology of changing forest stands on the western slope of the Cascade Mountains in Central Washington. M.S. thesis, Univ. Washington, Seattle.
Bowman, G. B., and L. D. Harris. 1980. Effect of spatial heterogeneity on ground-nest depredation. *J. Wildl. Manage.* 44:806–13.
Boyce, J. S. 1932. Decay and other losses in Douglas fir in western Oregon and Washington. U.S.D.A. Tech. Bull. 286.
Bratz, R. D. 1950. Avifaunal habitats in the central coast mountains of western Oregon. M.S. thesis, Oregon State Coll., Corvallis.
———. 1952. The distribution of biota in Oregon. Ph.D. diss., Oregon State Coll., Corvallis.
Brittingham, M. C., and S. A. Temple. 1983. Have cowbirds caused forest songbirds to decline? *BioScience* 33(1):31–35.
Brodie, J. D., R. O. McMahon, and W. H. Gavelis. 1978. Oregon's forest

resources: Their contribution in the state's economy. Res. Bull. 23. Corvallis: For. Res. Lab., Oregon State Univ.

Brown, J. H. 1971. Mammals on mountain tops: Nonequilibrium insular biogeography. *Am. Nat.* 105:467–78.

———. 1978. The theory of insular biogeography and the distribution of boreal birds and mammals. *Great Basin Nat. Mem.* 2:209–27.

Buckner, J. L., and J. L. Landers. 1980. A forester's guide to wildlife management in southern industrial pine forests. Bainbridge, Ga.: International Paper Co., Tech. Bull. 10.

Buechner, H. K. 1953. Some biotic changes in the state of Washington, particularly during the century 1853–1953. State Coll. Washington, *Res. Studies* 21:154–92.

Burgess, R. L., and D. M. Sharpe, eds. 1981. Forest island dynamics in man-dominated landscapes. New York: Springer-Verlag.

Burke, C. J. 1979. Historic fires in the central western Cascades, Oregon. M.S. thesis, Oregon State Univ., Corvallis.

Burns, R. 1973. Cultural change, resource use and the forest landscape: The case of the Willamette National Forest. Ph.D. diss., Univ. Oregon, Eugene.

Cain, S. 1938. The species-area curve. *Am. Midl. Nat.* 19:573–81.

Campbell, R. K. 1979. Genecology of Douglas fir in a watershed in the Oregon Cascades. *Ecology* 60:1036–50.

Clawson, M. 1976. The national forests. *Science* 191:762–67.

Connor, E., and E. McCoy. 1979. The statistics and biology of the species-area relation. *Am. Nat.* 113:791–33.

Cook, R. E. 1969. Variation in species density in North American birds. *Syst. Zool.* 18:63–84.

Council on Environmental Quality. 1980. The global 2000 report to the president, vol. 2. Council on Environmental Quality and the Department of State. G. O. Barney, study director. Washington, D.C.: U.S. Gov. Printing Office.

Cowan, I. McT. 1965. Conservation and man's environment. *Nature* 208:1145–51.

Crow, J. F., and M. Kimura. 1970. An introduction to population genetics theory. New York: Harper and Row.

Crowley, P. H. 1978. Effective size and the persistence of ecosystem. *Oecologia* 35:185–95.

Darling, F. F. 1952. Social behavior and survival. *Auk* 69:183–91.

Darlington, P. J., Jr. 1957. Zoogeography: The geographical distribution of animals. New York: John Wiley and Sons.

Demoulin, V., J. Duvigneaud, and J. Lambinon. 1976. Preserving national forests. Letters to the editor, *Science* 193:442.

Denison, W. C. 1973. Life in tall trees. *Scien. Am.* 228:74–80.

Denniston, C. 1977. Small population size and genetic diversity: Implications for endangered species. Pages 281–89 in S. A. Temple, ed., Endangered birds: Management techniques for preserving threatened species. Madison: Univ. Wisconsin Press.

Diamond, J. M. 1975. Assembly of species communities. Pages 342–444 in M. L. Cody and J. M. Diamond, eds., Ecology and evolution of communities. Pages 342–444 in M. L. Cody and J. M. Diamond, eds., Ecology and evolution of communities. Cambridge: Harvard Univ. Press.

Dirks-Edmunds, J. C. 1947. A comparison of biotic communities of the cedar-hemlock and oak-hickory associations. *Ecol. Monogr.* 17:235–60.

Dubos, R. 1970. The genius of the place. H. M. Albright Conservation lecture. Univ. Calif., Berkeley.

———. 1973. Humanizing the earth. *Science* 179:769–72.

Dyrness, C. T., J. F. Franklin, and W. H. Moir. 1974. A preliminary classification of forest communities in the central portion of the western Cascades in Oregon. Bull. 4. Seattle: Coniferous For. Biome, U.S.I.B.P., Univ. Washington.

Eckholm, E. 1980. Introduction. Pages 1–9 in Firewood Crops, shrub and tree species for energy production. Washington, D.C.: U.S. Natl. Acad. Sciences.

Edmonds, R. L. 1979. Western coniferous forests: How forest management has changed them. *Biol. Digest* 5:12–23.

Eltzroth, M. S., and F. L. Ramsey. 1979. Checklist of the birds of Oregon, 3rd ed. Corvallis: Audubon Soc. of Corvallis.

Evans, F. C., and P. J. Clark. 1954. Distance to nearest neighbor as a measure of spatial relationships in populations. *Ecology* 35:445–53.

Falconer, D. S. 1960. Introduction to quantitative genetics. New York: Ronald Press.

———. 1977. Why are mice the size they are? Pages 19–22 in E. Pollak, O. Kempthorne, and T. B. Bailey, eds., Proc. of the international conference on quantitative genetics. Ames: Iowa State Univ. Press.

———. 1981. Introduction to quantitative genetics. New York: Longman.

Farnum, P., R. Timmis, and J. L. Kulp. 1983. Biotechnology of forest yield. *Science* 219:694–702.

Fery, C. S. 1928. Timber appraisal in the Douglas fir region. M. For. thesis, Univ. Washington, Seattle.

Findley, J. S., and S. Anderson. 1956. Zoogeography of the mountain mammals of Colorado. *J. Mammal.* 37:80–82.

Ford, E. B. 1964. Ecological genetics. New York: John Wiley and Sons.

Forsman, E. D., E. C. Meslow, and M. J. Strub. 1977. Spotted owl abundance in young versus old-growth forests, Oregon. *Wildl. Soc. Bull.* 5:43–47.

Foster, R. 1980. Heterogeneity and disturbance in tropical vegetation. Pages 75–92 in M. E. Soulé and B. A. Wilcox, eds., Conservation Biology. Sunderland, Mass.: Sinauer Assoc. Inc.

Frankel, O. H. 1974. Genetic conservation: Our evolutionary responsibility. *Genetics* 78:53–65.

Frankel, O. H., and M. E. Soulé. 1981. Conservation and evolution. Cambridge: Cambridge University Press.

Franklin, I. R. 1980. Evolutionary change in small populations. Pages 135–49 in M. E. Soulé and B. A. Wilcox, eds., Conservation biology: An evolutionary-ecological perspective. Sunderland, Mass.: Sinauer Assoc. Inc.

Franklin, J. F. 1979. Vegetation of the Douglas-fir region. Pages 93–112 in P. E. Heilman, H. W. Anderson, and D. M. Baumgartner, eds., Forest soils of the Douglas-fir region. Pullman: Washington State Coop. Extn. Serv.

Franklin, J. F., K. Cromack, Jr., W. Denison, A. McKee, C. Maser, J. Sedell, F. Swanson, and G. Juday. 1981. Ecological characteristics of old-growth Douglas-fir forests. U.S.D.A. For. Serv., Gen. Tech. Rep. PNW-118.

Franklin, J. F., and C. T. Dyrness. 1973. Natural vegetation of Oregon and Washington. U.S.D.A. For. Serv., Gen. Tech. Rep. PNW-8.

Franklin, J. F., and R. H. Waring. 1980. Distinctive features of the northwestern coniferous forest: Development, structure, and function. Pages 59–86 in R. H. Waring, ed., Forests: Fresh perspectives from ecosystem analysis. Corvallis: Oregon State Univ. Press.

Fritschen, L. J., and P. Doraiswamy. 1973. Dew: An addition to the hydrological balance of Douglas fir. *Water Resour. Res.* 9:891–94.

Fritschen, L. J., C. H. Driver, C. Avery, J. Buffo, R. Edmonds, R. Kinerson, P. Schiess. 1971. Dispersion of air tracers into and within a forested area: 3. Res. and Dev. Tech. Rept. ECOM-68-G8-3, U.S. Army Elect. Command, Atmospheric Sci. Lab., Fort Huachuca, Arizona.

Gaines, D. A. 1980. The valley riparian forests of California: Their importance to bird populations. Pages 57–85 in A. Sands ed., Riparian forests in California, their ecology and conservation. Davis, Cal.: Priced Publ. 4101 Div. Agric. Sci. Univ. Calif.

Gannett, H. 1902. The forests of Oregon. Washington, D.C.: U.S.D.I. Geol. Survey. Prof. Pap. 4, Ser. H., For. 1.

Gashwiler, J. S. 1959. Small mammal study in west-central Oregon. *J. Mammal.* 40:128–39.

———. 1965. Tree seed abundance vs. deer mouse populations in Douglas-fir clearcuts. *Proc. Ann. Conf. Soc. Am. For.* 219–22.

———. 1967. Conifer seed survival in a western Oregon clearcut. *Ecology* 48:431–38.

———. 1970a. Plant and mammal changes on a clearcut in west-central Oregon. *Ecology* 51:1018–26.

———. 1970b. Further study of conifer seed survival in a western Oregon clearcut. *Ecology* 51:849–54.

———. 1972. Life history notes on the Oregon vole, *Microtus oregoni*. *J. Mammal.* 53:558–69.

Gavareski, C. A. 1976. Relation of park size and vegetation to urban bird populations in Seattle, Washington. *Condor* 78:375–82.

Gehrken, G. A. 1975. Travel corridor technique of wild turkey management. Pages 113–17 in L. K. Halls, ed., Proc. Natl. Wild Turkey Symp. 3. Austin: Texas Chapter, the Wildl. Soc.

Gilbert, F. S. 1980. The equilibrium theory of island biogeography: Fact or fiction? *J. Biogeog.* 7:209–35.

Gilbert, L. E. 1980. Food web organization and conservation of neotropical diversity. Pages 11–35 in M. E. Soulé and B. A. Wilcox, eds., Conservation biology: An evolutionary-ecological perspective. Sunderland, Mass.: Sinauer Assoc., Inc.

Gilbert, L. E., and P. H. Raven, eds. 1975. Coevolution of animals and plants. Austin: Univ. Texas Press.

Gilpin, M. E., and J. M. Diamond. 1980. Subdivision of nature reserves and the maintenance of species diversity. *Nature* 285–567–68.

Gittleman, J. L., and P. H. Harvey. 1982. Carnivore home-range size, metabolic needs and ecology. *Behav. Ecol. Sociobiol.* 10:57–63.

Gleason, H. 1922. On the relation between species and area. *Ecology* 3:158–62.

———. 1925. Species and area. *Ecology* 6:66–74.

Goertz, J. W. 1964. Habitats of three Oregon voles. *Ecology* 45:846–48.

Gordon, K. L. 1966. Mammals and the influence of the Columbia River gorge on their distribution. *Northwest Sci.* 40:142–46.

———. 1947. The origin and distribution of living North American mammals. Pages 25–31 in Proc. 8th Oregon State Coll. Biol. Colloquium: Biogeography. Corvallis.

Gosnell, M. 1976. The island dilemma. *Int. Wildl.* 6:24–35.

Grayson, D. K. 1982. Toward a history of Great Basin mammals during the past 15,000 years. Pages 82–101 in D. B. Madsen and J. F. O'Connell, eds., Society Am. Archaeology Pap. 2., Desert varnish: Man and environment in the Great Basin.

Greeley, W. B. 1943. A decade of progress in Douglas-fir forestry. Seattle: Joint Comm. For. Conserv., Pacific Northwest Loggers Assoc. and West Coast Lumbermen's Assoc.

Greig, J. C. 1979. Principles of genetic conservation in relation to wildlife management in southern Africa. *S. Afr. J. Wildl. Res.* 9:57–78.

Greig-Smith, P. 1957. Quantitative plant ecology. New York: Academic Press.

Grier, C. and R. Logan. 1977. Old-growth *Pseudotsuga menziesii* communities of a western Oregon watershed: Biomass distribution and production budgets. *Ecol. Monogr.* 47:373–400.

Grinnell, J. 1914a. An account of the mammals and birds of the lower

Colorado Valley with especial reference to the distributional problems presented. *Univ. California Publ. Zool.* 12:51–294.

———. 1914b. Barriers to distribution as regards birds and mammals. *Am. Nat.* 48:248–54.

———. 1916. An analysis of the vertebrate fauna of the Trinity region of northern California. *Univ. California Publ. Zool.* 12:399–410.

Grinnell, J., and H. S. Swarth. 1913. An account of the birds and mammals of the San Jacinto area of southern California, with remarks upon the behavior of geographic races on the margins of their habitats. *Univ. California Publ. Zool.* 10:197–406.

Groner, A. 1949. The use of plants as site indicators for Douglas-fir in Coos County, Oregon, and the correlation of site with soil properties. M. For. thesis, Univ. Washington, Seattle.

Gruson, E. S. 1976. Checklist of the world's birds. New York: Quadrangle.

Hamilton, G. A. 1962. Food web studies in an oak woodland ecosystem. M.S. thesis, Oregon State Univ., Corvallis.

Hansen, C. G. 1956. An ecological survey of the vertebrate animals on Steen's Mountain, Harney County, Oregon. Ph.D. diss., Oregon State Coll., Corvallis.

Harestad, A. S., and F. L. Bunnell. 1979. Home range and body weight—a re-evaluation. *Ecology* 60:389–402.

Harris, L. D. 1980. Forest and wildlife dynamics in the southeast. *Trans. North Am. Wildl. Nat. Resour. Conf.* 45:307–22.

Harris, L. D., and W. Smith. 1978. Relations of forest practices to non-timber resources and adjacent ecosystems. Pages 28–58 in T. Tippen, ed., Principles of maintaining productivity on prepared sites. New Orleans: U.S.D.A. For. Serv.

Harris, L. D., D. H. Hirth, and W. R. Marion. 1979. Development of silvicultural systems for wildlife. *Proceed. Ann. LA. St. Univ. Forestry Symp.* 28:65–81.

Harris, L. D., and P. Kangas. 1979. Designing future landscapes from principles of form and function. Pages 725–29 in G. H. Elsner and R. C. Smardon, tech. coord., Our national landscape. Proc. Conf. on applied techniques for analysis and management of the visual resource. U.S.D.A. For. Serv., Gen. Tech. Rep. PSW-35.

Harris, L. D., and P. Skoog. 1980. Wildlife habitat implications of forest management practices. In R. H. Chabreck and R. H. Mills, eds., Integrating timber and wildlife management in southern forests. *Proceed. Ann. LA. St. Univ. Forestry Symp.* 29:103–19.

Harris, L. D., and J. D. McElveen. 1981. Effect of forest edges on north Florida breeding birds. IMPAC Reports vol. 6, no. 4.

Harris, L. D., and W. R. Marion. 1981. Forest stand scheduling for wildlife in the multiple use forest. Pages 209–14 in Proc. of 1981 Convention of the Soc. of Am. Foresters.

Harris, L. D., C. Maser, and A. McKee. 1982. Patterns of old growth harvest and implications for Cascades wildlife. *Trans. North Am. Wildl. Nat. Resour. Conf.* 47:374–92.

Hartl, D. L. 1980. Principles of population genetics. Sunderland, Mass.: Sinauer Assoc.

Heaney, L. 1978. Island area and body size of insular mammals: Evidence from the tri-colored squirrel *(Callosciurus prevosti)* of Southeast Asia. *Evolution* 32:29–44.

Hempel, C. 1966. Philosophy of natural science. Englewood Cliffs: Prentice-Hall.

Hickman, J. C. 1968. Disjunction and endemism in the flora of the central western Cascades of Oregon: A historical and ecological approach to plant distributions. Ph.D. diss., Univ. Oregon, Eugene.

———. 1976. Non-forest vegetation of the central western Cascade mountains of Oregon. *Northwest Sci.* 50:145–55.

Honacki, J. H. 1978. Insular biogeography of montane mammals. *Trans. Kansas Acad. Sci.* 81:85.

Hooven, E. F. 1973. Response of the Oregon creeping vole to the clearcutting of a Douglas-fir forest. *Northwest Sci.* 47:256–64.

Hopkins, B. 1955. The species-area relations of plant communities. *J. Ecol.* 43:409–26.

Horton, R. E. 1945. Erosional development of streams and their drainage basins; hydrophysical approach to quantitative morphology. *Bull. Geol. Soc. Amer.* 56:275–370.

Hubbard, J. P. 1965. The summer birds of the forests of the Mogollon Mountains, New Mexico. *Condor* 67:404–15.

Hunt, G. L., Jr., and M. W. Hunt. 1974. Trophic levels and turnover rates: The avifauna of Santa Barbara Island, California. *Condor* 76:363–69.

Hutchinson, G. E. 1953. The concept of pattern in ecology. *Proc. Acad. Nat. Sci. Philadelphia* 105:1–12.

Hutchison, J. M., K. E. Thompson, and J. D. Fortune, Jr. 1966. The fish and wildlife resources of the upper Willamette basin, Oregon, and their water requirements. Portland: Basin Invest. Sect., Oregon State Game Comm.

Ingles, L. G. 1965. Mammals of the Pacific states. Stanford, Cal.: Stanford Univ. Press.

International Union for the Conservation of Nature and Natural Resources. 1980. World Conservation Strategy, living resource conservation for sustainable development. Gland, Switzerland: Prepared by the Intl. Union for Conserv. of Nature and Natural Resour. (IUCN), the United National Environ. Progr. (UNEP), and the World Wildl. Fund (WWF).

Isaac, L. A. 1940. Vegetative succession following logging in the Douglas-fir region with special reference to fire. *J. Forest.* 38:716–21.

———. 1952. Biological aspects of forest conservation in Washington and Oregon. Pages 12–15 in Proc. 13th Oregon State Coll. Biology Colloquium: Conservation. Corvallis.

Johns, P. E., R. Baccus, M. N. Manlove, J. E. Pinder, III, and M. H. Smith. 1979. Reproductive patterns, productivity and genetic variability in adjacent white-tailed deer populations. *Proc. Ann. Conf. S.E. Assoc. Fish and Wildlife Agencies* 31:167–72.

Johnson, M. 1973. Characters of the heather vole, *Phenacomys*, and the red tree vole, *Arborimus*. *J. Mammal.* 54:239–44.

Johnson, N. K. 1965. The breeding avifaunas of the Sheep and Spring Ranges in southern Nevada. *Condor* 67:93–124.

———. 1975. Controls of number of bird species on montane islands in the great basin. *Evolution* 29:545–67.

Kendeigh, S. C. 1944. Measurement of bird populations. *Ecol. Monogr.* 14:67–106.

Kiester, A. R. 1971. Species density of North American amphibians and reptiles. *Syst. Zool.* 20:127–31.

Kimura, M., and T. Ohta. 1971. Theoretical aspects of population genetics. Princeton: Princeton Univ. Press.

Kirkland, B. P. 1946. Forest resources of the Douglas-fir region. Portland: Joint Comm. For. Conserv., Pacific Northwest Loggers Assoc.

Kirkland, B. P., and J. F. Brandstrom. 1936. Selective timber management in the Douglas-fir region. Washington, D.C.: Div. For. Econ., U.S.D.A. For. Serv. Charles Lathrop Pack Forestry Foundation.

Koger, M. 1977. Summary. pages 434–47 in M. Koger, T. J. Cunha and A. C. Warnick, eds., Crossbreeding beef cattle, series 2. Gainesville: Univ. Florida Press.

Kubisiak, J. J., J. C. Moulton, and K. McCaffery. 1980. Ruffed grouse density and habitat relationships in Wisconsin. Madison, Wis.: Dept. Nat. Resour. Tech Bull. 118.

Lang, F. J. 1980. Stand age class survey of timber management areas in western Oregon and western Washington. Unpub. Rep. Prepared for Assoc. Oregon and California Counties. Sacramento: Jones and Stokes Assoc. Inc.

Langille, H. D., F. G. Plummer, A. Dodwell, T. F. Rixon, and J. B. Leiberg. 1903. Forest conditions in the Cascade Range Forest Reserve, Oregon. U.S.D.I. Geo. Survey, Series H, For. 6, Prof. Pap. 9.

Lanly, J. 1982. Tropical forest resources. F.A.O. Forestry paper 30. Rome: Food and Agriculture Organization of the United Nations.

Levenson, J. B. 1981. The southern mesic forest of southeastern Wisconsin: Species composition and community structure. Milwaukee: Contrib. Biol. Geol., Milwaukee Co. Public Museum.

Little, E. L., Jr. 1979. Checklist of United States trees native and naturalized. U.S.D.A. For. Serv., Agri. Handb. 541.

Lorenc, E. 1980. Analysis of fertility in inbred lines of mice derived from populations differing in genetic load. *Zwierzęta Laboratoryjne* 17:3–16.

Lovejoy, T. E. 1977. Genetic aspects of dwindling populations, a review. Pages 275–79 in S. A. Temple, ed., Endangered birds: Management techniques for preserving threatened species. Madison: Univ. Wisconsin Press.

Luman, I. D., and W. A. Neitro. 1980. Preservation of mature forest seral stages to provide wildlife habitat diversity. *Trans. North Am. Wildl. Nat. Resour. Conf.* 45:271–77.

MacArthur, R. H., and E. O. Wilson. 1963. An equilibrium theory of insular zoogeography. *Evolution* 17:373–87.

———. 1967. The theory of island biogeography. Princeton: Princeton Univ. Press.

MacClintock, L., R. Whitcomb, and B. Whitcomb. 1977. Evidence for the value of corridors and minimization of isolation in preservation of biotic diversity. *Am. Birds* 31:6–16.

Mace, G. M., and P. H. Harvey. 1983. Energetic constraints on home range size. *Amer. Natur.* 121:120–32.

Mannan. R. W. 1982. Bird populations and vegetation characteristics in managed and old-growth forests, northeastern Oregon. Ph.D. dissert. Oregon State Univ., Corvallis.

Martin, A. C., H. S. Zim, and A. L. Nelson. 1951. American wildlife and plants: A guide to wildlife food habits. New York: Dover.

Martin, T. E. 1978. Diversity and density of shelterbelt bird communities. Brookings: South Dakota State Univ.

———. 1981. Species-area slopes and coefficients: A caution on their interpretation. *Am. Nat.* 118:823–37.

Maser, C. O. 1966. Life histories and ecology of *Phenacomys albipes, Phenacomys longicaudus, Phenacomys silvicola*. M.S. thesis, Oregon State Univ.

Maser, C., R. Anderson, K. Cromack, Jr., J. T. Williams, R. E. Martin. 1979. Dead and down woody material. Pages 78–95 in J. W. Thomas, ed., Wildlife habitats in managed forests, the Blue Mountains of Oregon and Washington. Washington, D.C.: U.S.D.A. For. Serv., Agric. Handb. 553.

Maser, C., B. R. Mate, J. F. Franklin, and C. T. Dyrness. 1981. Natural history of Oregon coast mammals. U.S.D.A. For. Serv., Gen. Tech. Rep. PNW-133.

Maser, C., J. M. Trappe, and R. A. Nussbaum. 1978b. Fungal-small mammal inter-relationships with emphasis on Oregon coniferous forests. *Ecology* 59:799–809.

Maser, C., J. P. Trappe, and D. C. Ure. 1978a. Implications of small mammal mycophagy to the management of western coniferous forests. Trans. North Am. Wildl. Nat. Resour. Conf. 43:78-88.

May, R. M. 1975. Patterns of species abundance and diversity. Pages 81–120 in M. L. Cody and J. M. Diamond, eds., Ecology and evolution of communities. Cambridge: Harvard Univ. Press.

McKee, A., R. Nussbaum, and C. Maser. 1976. A checklist of the terrestrial vertebrates of the H. J. Andrews Experimental Forest. Unpubl. list.

McKeever, S. 1960. Food of the northern flying squirrel in northwestern California. *J. Mammal.* 41:270–71.

McNab, B. K. 1963. Bioenergetics and the determination of home range size. *Am. Nat.* 97:133–40.

Meadows, D. L. 1981. Coming wood energy demand on the forests. Pages 24–32 in 1981 Forest Industries Advisory Council Report and 1980 Forest Industries Council Annual Report.

Meslow, E. C. 1978. The relationship of birds to habitat structure—plant communities and successional stages. Pages 12–18 in R. M. DeGraaf, tech. coord., Proc. workshop on nongame bird habitat management in the coniferous forests of the western United States. U.S.D.A. For. Serv., Gen. Tec. Rept. PNW 64.

Meslow, E. C., C. Maser, and J. Verner. 1981. Old-growth forests as wildlife habitat. *Trans. North Am. Wildl. Nat. Resour. Conf.* 46:329–44.

Meslow, E. C., and H. M. Wight. 1975. Avifauna and succession in Douglas-fir forests of the Pacific Northwest. Pages 266–71 in D. R. Smith, tech. coord, Proc. Symp. Manage. of forest and range habitats for nongame birds. U.S.D.A. For. Serv., Gen. Tech. Rep. WO-1.

Meyers, N. 1980. Conversion of tropical moist forests. Washington, D.C.: U.S. Natl. Acad. Sciences.

Miller, A. 1955. The avifauna of the Sierra del Carmen of Coahuila, Mexico. *Condor* 57:154–78.

Miller, D. H. 1978. The factor of scale: Ecosystem, landscape mosaic, and region. Pages 63–88 in K. A. Hammond, G. Macinio, and W. B. Fairchild, eds., Sourcebook on the environment: A guide to the literature. Chicago: Univ. Chicago Press.

Miller, R. I. 1978. Applying island biogeographic theory to an East African reserve. *Environ. Conserv.* 5:191–95.

———. 1979. Conserving the genetic integrity of faunal populations and communities. *Environ. Conserv.* 6:297–304.

Miller, R. I., and L. D. Harris. 1977. Isolation and extirpations in wildlife reserves. *Biol. Conserv.* 12:311-15.

Muller, H. J. 1950. Our load of mutations. (Presidential address, read before American Society of Human Genetics, New York, December 28, 1949.) *Am. J. Human Genet.* 2:111–76.

Musiałek, B. 1980. Effect of the population size on decrease of fertility in mice. *Genetica Polonica* 21:461–475.

Nei, M., T. Maruyama, and R. Chakraborty. 1975. The bottleneck effect and genetic variability in populations. *Evolution* 29:1–10.

Olszewski, J. L. 1968. Role of uprooted trees in the movements of rodents in forests. *Oikos* 19:99–104.

Olterman, J. H., and B. J. Verts. 1972. Endangered plants and animals of Oregon. IV. Mammals. Agric. Exp. Sta. Spec. Rep. 364. Corvallis: Oregon State Univ.

Overton, W. S., and L. M. Hunt. 1974. A view of current forest policy, with questions regarding the future state of forests and criteria of management. *Trans. North Am. Wildl. Nat. Resour. Conf.* 39:334–53.

Patil, G. P., and C. Taillie. 1982. Diversity as a concept and its measurement. *J. Am. Stat. Assoc.* 77:548–61.

Paulson, J. T., and G. R. Leavengood. 1977. Timber management plan, Willamette National Forest. Portland: U.S.D.A. For. Serv., Region 6. With subsequent annual updates.

Pickett, S. T. A., and J. N. Thompson. 1978. Patch dynamics and the design of nature reserves. *Biol. Conserv.* 13:27–37.

Picton, H. D. 1979. The application of insular biogeographic theory to the conservation of large mammals in the northern Rocky Mountains. *Biol. Conserv.* 15:73–79.

Pielou, E. C. 1979. Biogeography. New York: John Wiley and Sons.

Pike, L. H., R. A. Rydell, and W. C. Denison. 1977. A 400-year-old Douglas-fir tree and its epiphytes: Biomass, surface area, and their distributions. *Can J. For. Res.* 7:680–99.

Plummer, F. G. 1902. Forest conditions in the Cascade Range, Washington, between the Washington and Mount Rainier Forest Reserves. U.S.D.I. Geol. Survey, Series H, For. 3, Prof. Pap. 6.

Portman, A. 1969. Animal forms and patterns. New York: Schocken Books.

Preston, F. W. 1962a. The canonical distribution of commonness and rarity: part I. *Ecology* 43:185–215.

———. 1962b. The canonical distribution of commonness and rarity: part II. *Ecology* 43:410–32.

Ralls, K., K. Brugger, and J. Ballou. 1979. Inbreeding and juvenile mortality in small populations of ungulates. *Science* 206:1101–3.

Ranney, J. W., M. C. Bruner and J. B. Levenson. 1981. The importance of edge in the structure and dynamics of forest lands. Pages 67–93 in R. L. Burgess and D. M. Sharpe, eds., Forest island dynamics in man-dominated landscapes. New York: Springer-Verlag.

Rejmanek, M. 1976. Centres of species diversity and centers of species diversification. *Evol. Biol. Praha.* 393–408.

Robbins, C. S. 1979. Effect of forest fragmentation on bird communities. Pages 198–212 in R. M. DeGraaf and K. E. Evans, eds., Management of north central and northeastern forests for non-game birds. Proc. Workshop U.S.D.A. For. Serv., Gen. Tech. Rep. NC-51.

Robbins, C. S., B. Bruun, and H. S. Singer. 1966. A guide to field identification: Birds of North America. Racine, Wis.: Western Publishing Co.

Robertson, A. 1952. The effect of inbreeding on the variation due to recessive genes. *Genetics* 37:189–203.

Rochelle, J. A. 1980. Mature forests, litterfall and patterns of forage quality as factors in the nutrition of black-tailed deer on northern vancouver island. Ph.D. diss., Univ. British Columbia.

Rudis, V. A., and A. R. Ek. 1981. Optimization of forest island spatial patterns: Methodology for analysis of landscape pattern. Pages 243–56 in R. L. Burgess and D. M. Sharpe, eds., Forest island dynamics in man-dominated landscapes. New York: Springer-Verlag.

Sargent, C. S. 1984. Forests of North America. 26 vols. U.S. Dept. Interior, census office. Report deriving from the 1880 census. Washington, D.C.: U.S. Gov. printing office.

Sauer, J. D. 1969. Oceanic islands and biogeographical theory: A review. *Geogr. Rev.* 59:582–93.

Schaaf, V. D. 1979. Threatened and endangered plant project report for the Willamette National Forest. Mimeo report.

Schamberger, M. L. 1970. The mammals of Mount Rainier National Park. Ph.D. diss., Oregon State Univ., Corvallis.

Schoen, J. W., O. C. Wallmo, and M. D. Kirchhoff. 1981. Wildlife-forest relationships: Is a re-evaluation of old-growth necessary? *Trans. North Am. Wild. Nat. Resour. Conf.* 46:531–44.

Senner, J. W. 1980. Inbreeding depression and the survival of zoo populations. Pages 209–24 in M. E. Soulé and B. A. Wilcox, eds., Conservation biology: An evolutionary-ecological perspective. Sunderland, Mass.: Sinauer Assoc., Inc.

Shaffer, M. L. 1981. Minimum population sizes for species conservation. *BioScience* 31:131–34.

Shaw, J. H., and P. A. Jordan. 1977. The wolf that lost its genes. *Natural History* 86(10):80–88.

Shelford, V. E., and E. D. Towler. 1925. Animal communities of the San Juan Channel and adjacent areas. *Publ. Univ. Washington, Puget Sound Bio. Stan.* 5:33–73.

Short, J. J. 1979. Patterns of alpha-diversity and abundance in breeding bird communities across North America. *Condor* 81:21–27.

Silen, R. R., and N. L. Mandel. 1982. Clinal genetic variation within Douglas-fir breeding zones—12 and 10 year progeny test results from two cooperative tree improvement programs. *J. For.* 81:216–20.

Simberloff, D. S. 1982. Island biogeographic theory and the design of wildlife refuges. *Ekologiya* (in press).

Simberloff, D. S., and L. G. Abele. 1982. Refuge design and island biogeography theory: Effects of fragmentation. *Am. Nat.* 120:41–40.

Simpson, G. G. 1964. Species density of North American recent mammals. *Syst. Zool.* 13:57–73.

Sirmon, J. M. 1982. Management aspects of old growth forest. In D. Johnson, ed., Old-growth forests, a balanced perspective. Proc. Conf. Feb. 12–14, 1982. Eugene, OR: Univ. Oregon Bur. Gov. Res., Eugene (in press).

Slatis, H. M. 1960. An analysis of inbreeding in the European bison. *Genetics* 45:275–87.

Smith, A. T. 1974. The distribution and dispersal of pikas: Consequences of insular population structure. *Ecology* 55:1112–19.

Smith, F. E. 1972. Spatial heterogeneity, stability, and diversity in ecosystems. *Trans. Connecticut Acad. Arts Sci.* 44:309–55.

Smith, M. H., C. T. Garten, Jr., and P. R. Ramsey. 1975. Genetic heterozygosity and population dynamics in small mammals. Pages 85–102 in C. L. Markert, ed., Isozymes IV: Genetics and evolution. New York: Academic Press.

Smith, R. H. 1979. On selection for inbreeding in polygynous animals. *Heredity* 43:205–11.

Snyder, N. F. R., and J. W. Wiley. 1976. Sexual size dimorphism in hawks and owls of North America. *Ornith. Monogr.* 26.

Sork, V. L. 1979. Demographic consequences of mammalian seed dispersal for pignut hickory. Ph.D. diss., Univ. Michigan, Ann Arbor.

Soulé, M. E. 1980. Thresholds for survival: Maintaining fitness and evolutionary potential. Pages 151–69 in M. E. Soulé and B. A. Wilcox, eds., Conservation biology: An evolutionary-ecological perspective. Sunderland, Mass.: Sinauer Assoc., Inc.

Stebbins, R. C. 1966. A field guide to western reptiles and amphibians. Boston: Houghton Mifflin.

Stevenson, H. M. 1976. Vertebrates of Florida. Gainesville: Univ. Presses of Florida.

Stiles, E. W. 1973. Bird community structure in alder forests. Ph.D. diss., Univ. Washington, Seattle.

———. 1980. Patterns of fruit presentation and seed dispersal in bird-disseminated woody plants in the eastern deciduous forest. *Am. Nat.* 116:670-88.

Strahler, A. N. 1957. Quantitative analysis of watershed geomorphology. *Trans. Am. Geophysical Union* 38:913–20.

Sturges, F. W. 1955. Habitats and distributions of land vertebrates on the Corvallis watershed, Mary's Peak, Benton County, Oregon. M. S. thesis, Oregon State Coll., Corvallis.

Sugihara, G. 1981. $S = CA^z, Z \simeq 1/4$: A reply to Connor and McCoy. *Am. Nat.* 117:790–93.

Sutton, E. H. 1966. Human heredity and its cytological bases. Pages 46–65 in J. B. Stanbury, J. B. Wyngaarden, and D. S. Frederickson, eds., The metabolic basis of inherited disease. New York: McGraw-Hill.

Swan, L. W. 1968. Alpine and aeolian regions of the world. Pages 29–54 in

H. E. Wright, Jr., and W. H. Osborn, eds., Proc. 7th Congr. Int. Assoc. Quat. Res. vol. 10.

Swanson, F. J. 1980. Geomorphology and ecosystems. Pages 159–70 in R. H. Waring, ed. Forests: Fresh perspectives from ecosystem analysis. Corvallis: Oregon State Univ. Press.

———. 1982. Fire and geomorphic processes. In H. A. Moody, M. Bonnicksen, N. L. Christensen, J. E. Lotan, and W. A. Reiners, eds., Fire regime and ecosystem properties. U.S.D.A. For. Serv., Gen. Tech. Rep. (in press).

Tatschl, J. L. 1967. Breeding birds of the Sandia Mountains and their ecological distributions. *Condor* 69:479–90.

Taylor, W. P., and W. T. Shaw. 1927. Mammals and birds of Mount Rainier National Park. Washington, D.C.: U.S. Gov. Printing Off.

Tedder, P. L. 1979. Oregon's future timber harvest: The size of things to come. *J. For.* 77:714–16.

Tesch, S. D. 1975. The composition and dynamics of the tree strata within an old growth Douglas-fir forest in western Montana. M.S. thesis, Univ. Montana, Missoula.

The Nature Conservancy. 1975. The preservation of natural diversity: A survey and recommendations. Report to the U.S. Dept. Interior by The Nature Conservancy. Washington, D.C.

Thomas, J. W., tech. ed. 1979. Wildlife habitats in managed forests, the Blue Mountains of Oregon and Washington. U.S.D.A. For. Serv., Agric. Handb. 553, Washington, DC. 512 pp.

Thomas, J. W., C. Maser and J. Rodiek. 1978. Edges—their interspersion, resulting diversity and its management. Pages 91–100 in R. DeGraaf, ed., Proc. workshop on nongame bird habitat management in the coniferous forests of the western United States. U.S.D.A. For. Serv., Gen. Tech. Rep. PNW 64.

———. 1979b. Edges. Pages 49–59 in J. W. Thomas, ed., Wildlife habitats in managed forests, the Blue Mountains of Oregon and Washington. Washington, D.C.: U.S.D.A. For. Serv., Agric. Handb. 553.

Thomas, J. W., R. G. Anderson, C. Maser, and E. L. Bull. 1979a. Snags. Pages 60–77 in J. W. Thomas, ed., Wildlife habitats in managed forests, the Blue Mountains of Oregon and Washington. Washington, D.C.: U.S.D.A. For. Serv., Agric. Handb. 553.

Thomas, J. W., G. L. Crouch, R. S. Bumstead, and L. D. Bryant. 1975. Silvicultural options and habitat values in coniferous forests. Pages 272–87 in D. R. Smith, tech. coord., Proc., Symp. on managment of forest and range habitats for nongame birds. Washington, D.C.: U.S.D.A. For. Serv., Gen. Tech. Rept. WO-1.

Thompson, D'Arcy. 1961. On growth and form. Abridged edition, J. T. Bonner, ed. Cambridge: Cambridge Univ. Press.

Thompson, L. S. 1974. Insular distribution of montane birds in the Sweetgrass Hills, Montana and other island mountain ranges of the northern Great Plains. M.S. thesis, Washington State Univ., Pullman.

Trappe, J., and R. Fogel. 1978. Ecosystematic functions of mycorrhizae. Pages 205–14 in J. Marshall, ed., The below ground symposium: A synthesis of plant associated processes. Range Sci. Dept. Ser. 26. Ft. Collins: Colorado State Univ.

Trappe, J., and C. Maser. 1977. Ectomycorrhizal fungi: Interactions of mushrooms and truffles with beasts and trees. Pages 165–79 in T. Walters, ed., Mushrooms and man, an interdisciplinary approach to mycology. Washington, D.C.: U.S.D.A. For. Serv.

Twight, P. A. 1973. Ecological forestry for the Douglas-fir region. Natl. Parks and Conserv. Assoc.

U.S. Department of Agriculture. Forest Service. 1978. Forest statistics of the U.S., 1977, Review Draft. Washington, D.C.: U.S. Gov. Printing Off.

———. 1980. 1978 wildfire statistics. FS 343.

U.S. Department of Agriculture. 1982. An analysis of the timber situation in the United States 1952-2030. U.S.D.A. Forest Service. Forest Resource Rept. 23.

U.S. Department of State. 1982. Proceedings of the U.S. strategy conference on biological diversity. Dept. State Publ. 9262, Intl. Organiz. and Conf. Series. 300. Washington, D.C.: U.S. Gov. Printing Off.

Vannote, R. L., G. W. Minshall, K. W. Cummins, J. R. Sedell, and C. E. Cushing. 1980. The river continuum concept. *Can. J. Fish. Aquat. Sci.* 37:130–37.

Veblen, T. T. 1982. Natural hazards and forest resources in the Andes of south-central Chile. *GeoJournal* 6:141–50.

Vestal, A. G. 1949. Minimum areas for different vegetations, their determination from species area curves. Urbana: Univ. Illinois Press.

Vestal, A. G., and M. Heermans. 1945. Size requirements for reference areas in mixed forest. *Ecology* 26:122–34.

Voth, E. 1963. A survey of the vertebrate animals of Mount Jefferson, Oregon. Ph.D. diss., Oregon State Univ., Corvallis.

Voth, E., C. Maser, and M. Johnson. 1983. Food habits of *Aborimus albipes*, the white-footed vole, in Oregon. *Northwest Sci.* 57:1–7.

Wallace, B. 1970. Genetic load, its biological and conceptual aspects. Englewood Cliffs, N.J.: Prentice-Hall.

Waring, R. H., and J. F. Franklin. 1979. Evergreen coniferous forests of the Pacific Northwest. *Science* 204:1380–86.

Weisbrod, A. R. 1976. Insularity and mammal species number in two national parks. Pages 83–87 in R. M. Linn, ed., Proc. 1st Conf. Sci. Res. Nat. Parks. U.S. Nat. Park Serv., Trans. and Proc. Series 5. 2 vols.

Westhoff, Dr. V. 1970. New criteria tor nature reserves. *New Scientist* 46(697):108–13.

Whitaker, J. O., Jr., and C. Maser. 1976. Food habits of five western Oregon shrews. *Northwest Sci.* 50:102–7.

Whitaker, J. O., Jr., C. Maser., and R. J. Pedersen. 1979. Food and ectoparasitic mites of Oregon moles. *Northwest Sci.* 53:268–73.

Whittaker, R. H. 1960. Vegetation of the Siskiyou Mountains, Oregon and California. *Ecol. Monogr.* 30:279–338.

Whittaker, R. H., and D. Goodman. 1979. Classifying species according to their demographic strategy. I. Population fluctuations and environmental heterogeneity. *Am. Nat.* 113:185–200.

Wiens, J. A. 1978. Nongame bird communities in northwestern coniferous forests. Pages 19–31 in R. M. DeGraaf, tech. coord., Proc. workshop on non-game bird habitat management in the coniferous forests of the western United States. U.S.D.A. For. Serv., Gen. Tech. Rept. PNW 64.

Wiens, J. A., and R. A. Nussbaum. 1975. Model estimation of energy flow in northwestern coniferous forest bird communities. *Ecology* 56:547–61.

Wight, H. M. 1974. Nongame wildlife and forest management. Pages 27–38 in H. C. Black, ed., Proc. symp. wildlife and forest management in the Pacific Northwest. Corvallis: For. Res. Lab., Oregon State Univ.

Wilcox, B. A. 1980. Insular ecology and conservation. Pages 95–117 in M. E. Soulé and B. A. Wilcox, eds., Conservation biology, an evolutionary-ecological perspective. Sunderland, Mass.: Sinauer Associates, Inc.

Willson, M. F., and S. W. Carothers. 1979. Avifauna of habitat islands in the grand canyon. *Southwestern Nat.* 24:563–76.

Wright, S. 1969. Evolution and the genetics of populations. Vol. 2: The theory of gene frequencies. Chicago: Univ. Chicago Press.

———. 1970. Random drift and the shifting balance theory of evolution. Pages 1–31 in K. Kojima, ed., Mathematical topics in population genetics. New York: Springer-Verlag.

———. 1977. Evolution and the genetics of populations. Vol. 3: Experimental results and evolutionary deductions. Chicago: Univ. Chicago Press.

Yancey, R. 1949. Animals of the subalpine forest of the McKenzie pass area. M.A. thesis, Oregon State Univ., Corvallis.

Zobel, D. B., A. McKee, G. M. Hawk, and C. T. Dyrness. 1976. Relationships of environment to composition, structure, and diversity of forest communities of the central western Cascades of Oregon. *Ecol. Monogr.* 46:135–56.

Author Index

Species Index

Subject Index